AF303208

Seed Survival in Genebanks –

Genetic and Biochemical Aspects of Seed

Deterioration in Barley

Dissertation
zur Erlangung des Doktorgrades
der Fakultät für Agrarwissenschaften
der Georg-August-Universität Göttingen

vorgelegt von

Manuela Nagel

geboren in Quedlinburg

Göttingen, Juli 2011

angefertigt im Leibniz-Institut für Pflanzengenetik und
Kulturpflanzenforschung Gatersleben

Bibliografische Information der Deutschen Nationalbibliothek
Die Deutsche Nationalbibliothek verzeichnet diese Publikation in der
Deutschen Nationalbibliografie; detaillierte bibliografische Daten
sind im Internet über http://dnb.d-nb.de abrufbar.
1. Aufl. - Göttingen : Cuvillier, 2011
    Zugl.: Göttingen, Univ., Diss., 2011

    978-3-86955-871-4

Referent: Prof. Dr. Rolf Rauber (Universität Göttingen)

Korreferent: Priv. Doz. Dr. Andreas Börner (IPK Gatersleben)

Tag der mündlichen Prüfung: 21. Juli 2011

© CUVILLIER VERLAG, Göttingen 2011
    Nonnenstieg 8, 37075 Göttingen
    Telefon: 0551-54724-0
    Telefax: 0551-54724-21
    www.cuvillier.de

Structure

List of abbreviations

AA	Artificial ageing, accelerated ageing
ABA	Abscisic acid
ABI	Abscisic acid insensitive mutants
AFLP	Amplified fragment-length polymorphism
AFR	Africa
ANOVA	Analysis of variance
AP2	APETALA2
APS	Arabian Peninsula
APX	Ascorbate peroxidase
AS	Ambient storage ($20°C$)
AsA	Ascorbic acid, ascorbate
ATP	Adenosine-5'-triphosphate
BLASTX	Basic Local Alignment Tool
c	Centromere
CAT	Catalase
CBD	Convention on Biological Diversity
CD	Controlled deterioration
cM	CentiMorgen
CS	Cold Storage ($\leq 0°C$)
CYS	Cysteine
γ-CYS-GLY	γ-Cysteinylglycine
DArT	Diversity Array Technology
DH	Doubled haploid
DHA	Hydroascorbate
DNA	Deoxyribonucleic acid
DP	Dipeptidase
DREB	Dehydration-responsive element binding
γ-ECS	γ-Glutamylcysteine synthetase
$E_{GSSG/2GSH}$	Half-cell reduction potential of glutathione
ERE	Ethylene-responsive element
E:S	Embryo to seed ratio
EST	Expressed sequence tags
GA	Gibberellic acid
GBR	Gatersleben barley RFLP markers
GBM	Gatersleben barley microsatellite markers
GBS	Gatersleben barley SNP markers

II

GLM	General linear model
GLU	Glutamate
γ-GLU-CYS	γ-Glutamylcysteine
GLY	Glycine
GR	Glutathione reductase
GS	Glutathione synthase
GSH	Glutathione
GST	Glutathione *S*-transferase
GSSG	Glutathione disulfide
γ-GT	γ-Glutamyl transpeptidase
H_2O_2	Hydrogen peroxide
HOR	*Hordeum* sp., barley genebank accession
HPLC	High-performance liquid chromatography
HSP	Heat shock proteins
I	Initial
ICARDA	International Center for Agricultural Research in the Dry Areas
IPK	Leibniz-Institut für Pflanzengenetik und Kulturpflanzenforschung
ISTA	International Seed Testing Association
LD	Linkage disequilibrium
LEA	Late embryogenesis abundant
LiCl	Lithium chloride
LOD	Logarithm of odds
LRR	Leucine rich repeats
MAF	Minor allele frequency
MAPK	Mitose-activated protein kinase
MC	Moisture content
MCMC	Markov chain Monte Carlo approach
MD	Morphological dormancy
MEA	Middle East Asia
MLM	Mixed linear model
MPD	Morphophysiological dormancy
mRNA	Messenger ribonucleic acid
M08	Multiplication of ICARDA population in 2008
MTA	Marker-trait association
NADPH	Nicotinamide adenine dinucleotide phosphate
NBS	N-terminal nucleotide binding site
NCBI	National Center for Biotechnology Information
NEA	North East Asia
NMR	Nuclear magnetic resonance

NRPEP	Non - redundant protein database
smc	Seed moisture content, dry oven method according to ISTA (2008)
TKW	Thousand kernel weight
$O_2^{\cdot-}$	Superoxide radical ion
OH	Hydrogen bonding, interaction between hydrogen and oxygen
$OH^{\cdot}$	Hydroxyl radical
OWB	Oregon Wolfe Barley population
PCA	Principle Component Analysis
PCD	Programmed cell death
PCS	Phytochelatins
PCoA	Principle Coordinate Analysis
PD	Physiological dormancy
PD+PY	Combinational dormancy (PD+PY)
PUFA	Poly-unsaturated fatty acid
Px	Peroxidase
PY	Physical dormancy
Q	Subpopulations
QTL	Quantitative trait loci
RFLP	Restriction fragment length polymorphism
RH	Relative humidity (in %)
ROS	Reactive oxygen species (e.g. $OH^{\cdot}$, $O_2^{\cdot-}$, H_2O_2)
SFG	Samenfeuchtegehalt
SNP	Single-nucleotide polymorphism
SxM	Steptoe x Morex population
SOD	Superoxide dismutase
SSR	Simple Sequence Repeats, microsatellites
T3	Tocotrienols
T_g	Glass transition temperature (temperature at which a substance transforms from a glass into a liquid)
TC	Total carbon
TG	Total germination
TILLING	Targeting Induced Local Lesions IN Genomes
TN	Total nitrogen
TOC	Tocopherols
UPGMA	Unweighted pair group method with arithmetic average
Ψ	Water potential

1   Prolog

1.1 Orthodox seeds - survival in a dry state

1.1.1   Evolutional aspects of seeds

Unlike animals or humans plants cannot escape from their surroundings, therefore they have adapted themselves to their changing environments over thousands of years ago. One of the evolutionary most successful adaptation was the appearance of seed plants (*Spermatophyta*) which comprises the gymnosperms (*Acrogymnospermae*) with nowadays about 800 species and the angiosperms or flowering plants (*Angiospermae*) with about 250,000 species (Cantino *et al.* 2007, Linkies *et al.* 2010). Especially flowering plants seemed to evolve abruptly in the mid-cretaceous and spread in a highly accelerated rate of diversification which was named by Charles Darwin as 'an abominable mystery' (Friedman 2009). The reasons for this development are widely hypothesised but mainly accepted is the mutually beneficial animal-plant relationship including the remarkable co-evolution of the angiosperms and the pollination/dispersal animal agents. High speciation rates, among the annual growth form, homeotic gene effects, asexual/sexual reproduction, the propensity for hybrid polyploidy, as well as a low extinction rate, or broad ecological tolerances have to be shown empirically as additional advantages (Crepet & Niklas 2009).

The major points of the fast dominance of angiosperms are the seeds which are, in contrast to gymnosperms, enclosed inside an ovary. Seeds are the dispersal unit of the angiosperms and consist of one or more mature seeds in the same ovary which are protected by a pericarp or fruit coat that can contain additional flower parts (Linkies *et al.* 2010). A further differentiator is the double fertilization, that is, in addition to the egg cell fertilization, a second fertilization of a central cell nucleus which leads to the formation of diploid or triploid endosperm (Friedman 1995, Friedman & Ryerson 2009). It is assumed to enhance the fitness of the embryo, serves a nutrient source during seed development and is, in some cases, also involved in the control of germination by being a barrier for the growing radicle (Linkies *et al.* 2010).

Seed of about 90% of the angiosperms species is desiccation tolerant (Hoekstra 2005, Alpert 2006) and termed 'orthodox' (Colville & Kranner 2010). This is remarkable because these organisms are able to tolerate an absolute water content of less than 10% and regain their normal function after rehydration. The achieved water content is equivalent to equilibration with air of 50% relative humidity at 20°C or to dropping to a water potential of -100 MPa and corresponds to a point at which enzymatic reactions and thus metabolism probably stops (Alpert 2005). In this state organisms are able to overcome periods of unfavourable conditions such as drought, high-energy radiation or prolonged exposure to very low temperatures (França *et al.* 2007). In contrast, desiccation sensitive seeds which are termed 'recalcitrant' do not undergo maturation drying and remain metabolically active after they are shed from the parent plant. These kind of seeds are adapted to germinate immediately and cease in viability as soon as the water content declines (Berjak & Pammenter 2008).

The ability of seeds to remain viable after dehydration is thought to be an ancient trait. It was lost during the evolution of structural and morphological modifications of vegetative tissue in association with internal water transport (Alpert 2005, Illing *et al.* 2005). The subsequent successful distribution of flowering plants was probably a consequence of the evolution of desiccation tolerance in seeds, in parallel, which re-evolved desiccation tolerance in vegetative tissue again and allowed greater control of water status (Oliver *et al.* 2000).

The anhydrobiotic character which identifies desiccation tolerant species enables most seed to extend their life span from months to decades under ambient conditions (Priestley *et al.* 1985, Nagel & Börner 2010) and can even exceed to 2,000 years (Sallon *et al.* 2008) as shown in Tab. 1. However, the absolute seed life spans are the results of environmental conditions in combination with intrinsic properties which we want to refer to during the next chapters.

Tab. 1: Exceptional seed longevities discovered during the last two decades.

Longevity (years)	Species	Place of discovery	Source
2,000	Date palm (*Phoenix dactylifera* L.)	Herodian fortress	Sallon *et al.* (2008)
1,300	Lotus (*Nelumbo nucifera* Gaertn.)	Chinese dry lakebed	Shen-Miller *et al.* (1995)
200	*Liparia* sp., *Acacia* sp., *Leucospermum* sp.	British Archives/ Tower of London	Daws *et al.* (2007)
150	*Acacia* sp.	Swedish Museum	Leino & Edqvist (2010)
110	Barley (*Hordeum vulgare* L.), oat (*Avena sativa* L.), weed seeds	Haberland's experiment in Vienna	Steiner & Ruckenbauer (1995)

1.1.2 Glassy state

The ability of orthodox seed to resist complete dehydration is due to dramatic decrease of molecular mobility and the transition of cytoplasm to the glassy or vitreous state (Sun & Leopold 1997, Buitink & Leprince 2004). A glass is an amorphous, solid state with an extremely high viscosity which is comparable to crystalline materials having defined structures, stoichiometric compositions and melting points. At the same time it is also an amorphous matrix including random position of molecules with restricted mobility like a super-cooled liquid (Sun 1997, Buitink *et al.* 1998, Buitink & Leprince 2004) equipped with 'holes' and open spaces which enable small molecules to diffuse through (Walters 1998). Its flow rate is in the order of 10^{-14} m·s^{-1} compared to a typical liquid of 10 m·s^{-1} (Franks *et al.* 1991). A key feature of characterisation is the glass transition temperature (T_g) which describes the temperature when a substance transforms from glass into liquid (Buitink & Leprince 2004).

When water is removed, the packing density of the phospholipids head groups in membranes increases and results in stronger van der Waals forces among hydrocarbon chains (Crowe *et al.* 1992). Consequently, the phase transition temperature increases significantly and dry lipids turn into a gel phase at room temperature (Hoekstra 2005).

Carbohydrates are important players by forming sugar glasses which is related with their ability to replace the hydrogen (OH-) bonding function of water. This property enhances the maintenance of the native structure of membranes and biopolymers (Hoekstra 2005). The molecular packing of several mono-, di- and tri- saccharides, such as glucose, trehalose and raffinose showed that monosaccharide glasses exhibited a stronger OH-bonding network than glasses made of higher molecular weight sugars. These suggest that smaller monosaccharide units can be packed more tightly in the glassy state (Buitink & Leprince 2004, Wolkers *et al.* 2004, Hoekstra 2005). Especially the non-reducing disaccharide trehalose is one of the most effective osmoprotectant sugars (Crowe *et al.* 1992, Iturriaga *et al.* 2009). It is thought to prevent damage by replacing the water shell around molecules (Patist & Zoerb 2005) and due to its high T_g it leads to stable glass formation (Iturriaga *et al.* 2009). There is growing evidence that trehalose, additionally, acts as an antioxidant by reducing intracellular oxidation and lipid peroxidation during dehydration (Pereira *et al.* 2003).

In accordance with different studies it is accepted that intracellular glasses are not only made of sugar molecules alone. The phase diagram of glass transitions for embryo tissues is distinctively different from that of carbohydrate mixes (Sun & Leopold 1995, Buitink *et al.* 2000, França *et al.* 2007). The addition of peptides (Wolkers *et al.* 1998), Late Embryogenesis Abundant (LEA) proteins (Wolkers *et al.* 2001) or tricarboxylic acid salts (Kets *et al.* 2004) enhanced further the hydrogen bonding strength and increased the T_g. In particular, LEA proteins and also heat shock proteins (HSP) facilitate glass formation (Buitink & Leprince 2004) and act as molecular shields which prevent molecular interactions, membrane fusion and protein aggregations. Improved glass characteristics are only achieved in cytoplasmic glasses where compounds, such as amino acids, ions and salts contribute to a denser OH-bonding network (Hoekstra 2005). In this state the activity of enzymatic and structural integrity remains, the secondary structure of proteins appears to be very stable and deteriorative chemical reactions are slowed down over several years of storage (Buitink & Leprince 2004).

A direct relationship between intermolecular hydrogen bonding strength in the glassy cytoplasm or the T_g and the life span could not be proved (Hoekstra 2005) but a relationship between seed longevity and the glass transition kinetics itself has been hypothesised by Sun (1997). His study indicated that the rate of seed viability loss during storage was affected by the water content which was correlated with the plasticization effect of water on intracellular glasses. Conditions, as for example higher water content and temperature, bring anhydrobiotes above their T_g and result in a linear increase of activation energy and rotational mobility of molecules in the cytoplasm. Buitink *et al.* (2000) could demonstrate that the stability of *Arabidopsis* seeds correlates with the density of the molecular matrix.

Summarizing, there is the suggestion that ageing rate and thus life span of germplasm can be influenced by the molecular stability of the cytoplasm as an important function of intracellular glasses (Buitink & Leprince 2004). However, the lack of direct correlations between OH-bonding strength or T_g might indicate that the glassy cytoplasm is not the system that is the most sensitive to ageing (Hoekstra 2005).

1.1.3 Oxidative stress and seed protection mechanisms

The extension or reduction of seed life span has already been mentioned to be related to the glassy state influenced by the effect of water and temperature. According to Harrington's (1963) thumb rule life span is doubled by decreasing seed moisture content (smc) by one percent or cooling by 5°C. Due to the high importance of water, Walters *et al.* (2005b) categorised seed ageing processes into five hydration levels that correspond to critical water (Ψ) potential and relative humidity (RH) as shown in Fig. 1.

Hydration levels

Seeds above a water potential $\Psi \geq -1$ MPa are fully hydrated, cell division is initialised and they start to grow (Hydration Level 5). Between $-1 \geq \Psi \geq -3$ MPa (Hydration Level 4) stress response and repair mechanism are detectable which are particular advantageous for seed priming. At Hydration Level 3 ($-5 \geq \Psi \geq -15$ MPa) intensive radical production and loss of membrane integrity occurs. When the water potential decreases to less than -15 MPa, which is comparable with a water content of about 0.10 g $H_2O \cdot$(g dry mass $-$ g lipid)$^{-1}$ at $\approx 50\%$ RH and $\leq 25°C$, seeds enter the glassy state (Hydration Level 2) and longevity extends. At a critical point ($\Psi \approx -200$ MPa, Hydration Level 1) a switch to degrading reactions including decreasing longevity is indicated (Walters *et al.* 2005b).

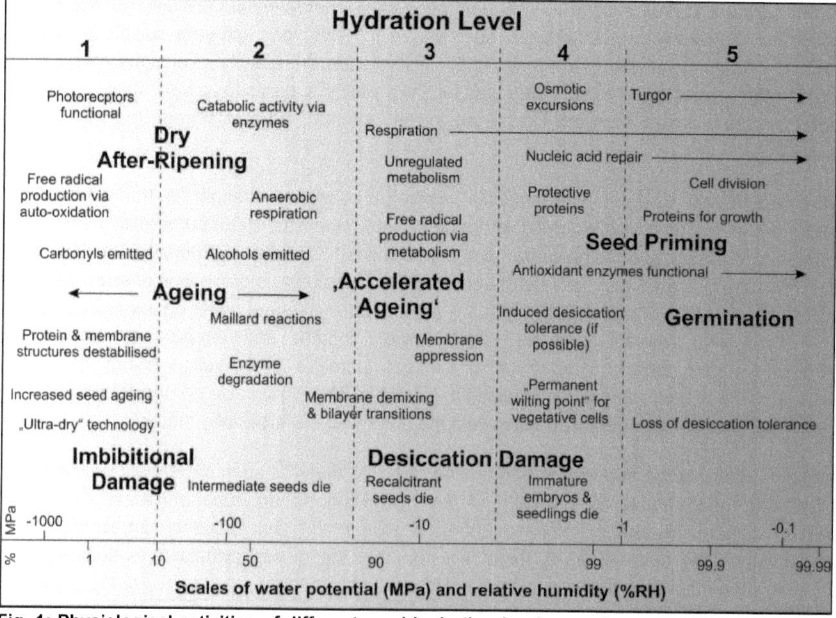

Fig. 1: Physiological activities of different seed hydration levels as a function of water potential and relative humidity. Data are adapted from Walters *et al.* (2005b) according to information of Vertucci & Roos (1990), Vertucci & Farrant (1995), Walters (1998) and unpublished data.

Stress and protection at Hydration Level 2 and lower

At Hydration Level 2 and lower seeds have entered the glassy state and metabolic processes are unlikely, although metabolic reactions cannot be totally excluded and might continue at very slow rates (Bailly *et al.* 2008, Kranner *et al.* 2010b). Within this amorphous matrix the speed of deleterious chemical processes is significantly slowed down (Sun & Leopold 1995). However, small molecules are able to penetrate holes and open space of the glassy environment and cause irreversible degradations (Walters 1998) which are even traceable in ancient barley (*Hordeum sp.*) kernels and radish (*Raphanus sativum* L.) seed (Bergen *et al.* 1997, Evershed *et al.* 1997).

These molecules are mainly reactive oxygen species (ROS) originating during abiotic and biotic stresses that enhance an impairment of the photosynthetic and respiratory electron transport (Halliwell 2006, De Gara *et al.* 2010). They encompass a variety of diverse chemical species including extremely unstable superoxide anions ($O_2^{\cdot-}$) and hydroxyl radicals ($OH^{\cdot}$) and the free diffusible and relatively long-lived hydrogen peroxide (H_2O_2) (Finkel & Holbrook 2000). Although the photosynthetic activity declines with maturation in orthodox seed (El-Maarouf-Bouteau & Bailly 2008) the metabolic ROS formation might be confined but cannot be entirely avoided and will occur through autoxidation (Colville & Kranner 2010). It is also expected that dehydration in the presence of light will accumulate free radicals (Khan *et al.* 1996).

Lipids and poly-unsaturated fatty acids (PUFAs) are one of the major targets of free radicals which often cause an extensive peroxidation (lipid peroxidation) and de-esterification of membrane lipids resulting in perturbation of membranes and water loss (Senaratna *et al.* 1987). A similar kind of molecular rearrangement under dry conditions is achieved by the reaction of carboxyl groups of sugars or aldehydes with amino groups of proteins forming complex mixtures (Amadori and Maillard reaction) which are responsible for odours and flavours (Walters 1998, Strelec *et al.* 2008). Proteins are in general targets of reactive radicals and scavenge between 50 and 70% ROS (Davies *et al.* 1999). If proteins are modified they will be possibly irreversible changed in their tertiary structure resulting in degradations and their progressive accumulation (Colville & Kranner 2010). Kibinza *et al.* (2006) showed that the occurrence of lipid peroxidation or oxidative damage of non lipid cellular fractions (proteins, nucleic acids) are dependent on the availability of water.

Apart from proteins, damage to seed nucleic acids including DNA single-strand break caused by direct ROS attack of deoxyribose units (Bray & West 2005) will change DNA content (Sen & Osborne 1974, 1977) and lead to DNA fragmentation (Osborne 2000, Kranner *et al.* 2011). Double strand breaks result in a loss of genetic information if homologous recombination and non-homologous end-joining repair pathways are not initiated (Bray & West 2005, Waterworth *et al.* 2007). However, the DNA stability during dehydration is an important feature of desiccated seeds (Boubriak *et al.* 1997) and mainly safeguarded by a ROS scavenging system. Catalase (CAT2) and cytosolic ascorbate peroxidase (APX1) play a key role against photorespiratory-dependent H_2O_2 induced DNA damage in hydrated plant tissue (Vanderauwera *et al.* 2011).

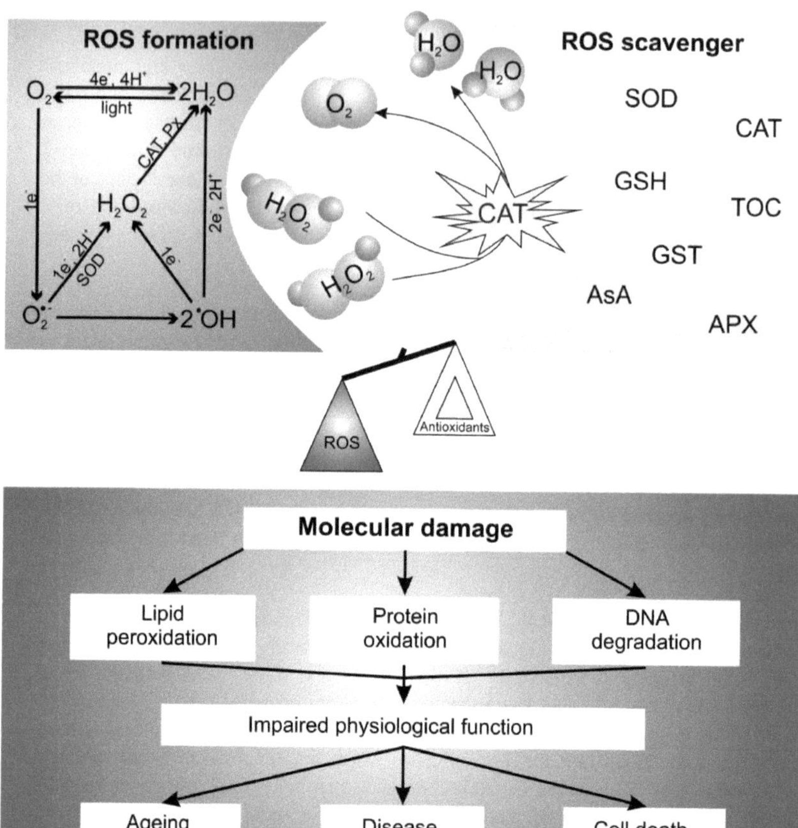

Fig. 2: Sources and cellular responses to reactive oxygen species (ROS). Oxidants are generated as a result of enzymatic and non-enzymatic reactions and are regulated by antioxidant defence systems, including catalase (CAT), superoxide dismutase (SOD), peroxidase (Px), glutathione (GSH), tocopherols (TOC), ascorbate (AsA), glutathione *S*-transferase (GST) and AsA peroxidase (APX). Figure is adapted from Franca *et al.* (2007).

In orthodox seed, protection against oxidative damage is ensured by a complex defence system (Fig. 2) that includes enzymes, such as peroxidases, catalases and superoxide dismutases (Sung & Jeng 1994, Kibinza *et al.* 2006, Lehner *et al.* 2008, Cakmak *et al.* 2010) and low-molecular-weight antioxidants, such as ascorbate (AsA), glutathione (GSH) and tocopherols (Kranner *et al.* 2010b). The major antioxidants AsA and GSH are water soluble and scavenge free radicals directly by acting as electron donors for ROS-detoxifying enzymes as AsA peroxidase (APX) and glutathione *S*-transferase (GST). Further both antioxidants participate in the ascorbate–glutathione cycle where GSH is needed for the reduction of dehydroascorbate (DHA) by the action of ascorbate peroxidase or by non-enzymatic reactions of ascorbate with oxidants. In these enzymatic or chemical reactions, GSH is oxidised to glutathione disulphide (GSSG) and reduced by glutathione reductase

(GR) at the expense of NADPH (Kranner *et al.* 2010b, Foyer & Noctor 2011). In the dry state AsA and GSH cannot be regenerated by reaction with ROS and the AsA/ DHA pool becomes even depleted because of the unstable DHA which is non-enzymatically broken down (De Tullio & Arrigoni 2003). GSSG on the other hand accumulates and can react with thiol groups of proteins forming protein-SSG mixed disulphides, which protect proteins against desiccation-induced oxidative injuries (Kranner & Grill 1996). The ratio between glutathione and the disulphide-glutathione (GSSG/2GSH) defines the most abundant redox couple in cells. Changes of its half-cell reduction potential $E_{GSSG/2GSH}$ appear to correlate with the biological status of the cells (Schafer & Buettner 2001) and are assumed to be part of the oxidative stress signalling cascade (Foyer & Noctor 2011) and programmed cell death (PCD) (Kranner *et al.* 2006).

Interestingly, signalling pathways are operative in desiccated seeds and are able to perceive environmental cues, such as those required for dormancy breaking (Finch-Savage *et al.* 2007). The mobility of protons might be enabled by the existence of hydrated pockets in which some metabolic activities are possible (Leubner-Metzger 2005). El-Maarouf-Bouteau & Bailly (2008) even considered that ROS could be sensed and, in particular, short-lived radicals such as ˙OH react with receptor proteins close to their production site (Moller *et al.* 2007).

Thus, as demonstrated in Fig. 2, dry seeds can withstand the attack of free radicals by a complex ROS scavenging system but ongoing oxidative stress can lead to an imbalance between prooxidant and antioxidant activities (Rajjou & Debeaujon 2008). During storage the detoxification potential is strongly altered by a continuous depletion of antioxidant enzymes, low molecular antioxidants and a change of $E_{GSSG/2GSH}$ (Colville & Kranner 2010). Consequently uncontrolled oxidative attacks cause damage to macromolecules, phospholipids, PUFAs and proteins (Kranner *et al.* 2010b) resulting in a loss of membrane integrity and death (Tammela *et al.* 2003, Pukacka & Ratajczak 2007).

Stress and protection at Hydration Level 3 and higher

When accumulated damage did not result in viability loss the detoxification system can be restored by priming treatment which is the initialisation of germination events by slow water uptake and a drying to its initial moisture content (Varier *et al.* 2010). The advantageous effect is due to the re-establishment of initial GSH concentration and $E_{GSSG/2GSH}$ by rapid reduction of GSSG (Kranner *et al.* 2005), the synthesis of antioxidant enzymes (Yeh & Sung 2008) and heat shock proteins which are assumed to enhance the folding of proteins and prevent the binding to damaged proteins (Lee *et al.* 1995). Reparation of DNA is mainly required for generation of error-free templates for transcription and replication and enzymes catalysing mobilization of storage reserve proteins. Comparable reactions occur during germination except for the faster water uptake which shortens the time for repair mechanism (Varier *et al.* 2010).

Simultaneously when imbibition starts also metabolic processes are initialised and mitochondria become source of ROS by producing hydrogen peroxide (H_2O_2) (Oracz *et al.* 2007). Although these ROS produce additional oxidative stress they are supposed to play a key role as messengers or transmitters of environmental cues (Mittler 2002). As signalling molecules they are involved in the later discussed transition from a dormant to a non-dormant seed (Bailly *et al.* 2008). They contribute to cell wall loosening during endosperm weakening (Müller *et al.* 2010) and defend the emerging seedling against pathogens together with roles in growth and development (Kranner *et al.* 2010c).

The role of temperature

Apart from the important impact that water is having on chemical reactions also the temperature influences the glass behaviour (Walters *et al.* 2005b) and affects kinetics of deteriorative reactions (Walters 2004). According to Harrington (1963) a decrease of temperature by 5°C doubles the life span of seeds. A rise in temperature leads to an increase of molecular motions and enhances the number of intermolecular collisions, which in turn enables the reaction to proceed. Near the phase transition zone, small changes in temperature can result in large viscosity changes (Walters 2004). For describing the relative effects of temperature on molecular mobility and structure of glasses the terms 'fragile' and 'strong' were defined by Angell (1991). Following Walters *et al.* (2005b) fragile glasses are affected by small changes in temperature where strong glasses require larger temperature changes for comparable disruption of glassy structure. Further combination of constants by Buitink & Leprince (1990), Vertucci & Ross (1990) and Walters (1998) lead to the suggestion that glasses are strong in seeds within Hydration Level 1 and fragile in seeds within Hydration Level 2. Latter case supports the idea that seed longevity can be strongly influenced in Hydration Level 2 (Walters *et al.* 2005b).

1.1.4 Programmed Cell Death (PCD)

Independently of water and temperature level, an excess accumulation of ROS and damage to macromolecules produce secondary messengers that play signalling roles in PCD pathways (Mittler *et al.* 2004, Bailly *et al.* 2008, Kranner *et al.* 2010b). The programmed cell death which occurs during development and in response to environmental cues is essential for survival of plants (Greenberg 1996). It allows organism to get rid of unwanted cells or cells that have been damaged beyond repair and enables the control of cellular developments and mechanism of biotic and abiotic stress defence (Hengartner 2000, Kranner *et al.* 2010b). PCD is a part of the resistance response and assures the survival of sufficient cells but if too many cells die, especially seed embryo cells, the whole seed will lose its viability (Osborne 2000, Kranner *et al.* 2006, Kranner *et al.* 2010b).

A part of the regulatory function of PCD is autophagy ('self-eating') which involves two modes in the dynamic rearrangement of discarded membranes: microautophagy and macroautophagy. The two modes differ with respect to the pathway by which cytoplasmic material is delivered to the lysosome but they share the final steps of lysosomal digestion (Levine & Klionsky 2004). Microautophagy which engulfs cytoplasm directly at the lysosomal surface (Levine & Klionsky 2004) is assumed to accompany seed germination (Toyooka *et al.* 2001). Starch granules and storage proteins are digested in *Vigna mungo* seed by lytic vacuols (Toyooka *et al.* 2001) in the same way as barley (*Hordeum vulgare* L.) aleuron layers are degraded after seedling establishment (Fath *et al.* 2001).

The pathways leading to PCD and autophagy are strongly induced by oxidative stress (Bassham 2007). As demonstrated in the previous chapter even in the glassy state these stress signals are transmitted through the cytoplasm (Finch-Savage *et al.* 2007) and could activate MAPK cascade, BAX- and BCL-like genes, Ca^{2+} signalling, and metacaspases (Saitoh *et al.* 1998, Mattson & Chan 2003, Riedl & Shi 2004) following the pathway in Fig. 3. In the final phase of the programmed cell death DNA is cleaved into intranucleosomal fragments which is described by Hengartner (2000) as hallmark of PCD. This kind of DNA fragmentation, visualised by "DNA laddering," occurs with increasing intensity as seeds lose viability and it is assumed to be the controlled pattern of PCD in low hydrated orthodox seed (Kranner *et al.* 2006).

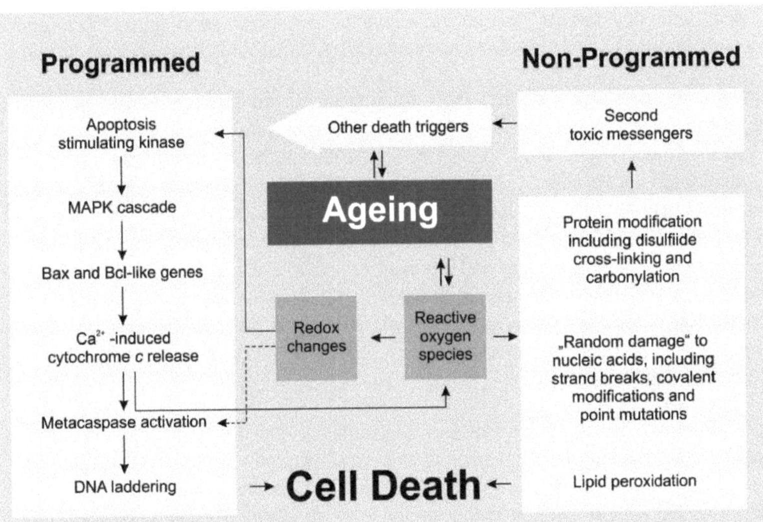

Fig. 3: **Simplified scheme of mechanisms that contribute to cell death during ageing (Kranner *et al.* 2011).** MAPK, mitogen-activated protein kinase, DNA, Deoxyribonucleic acid

Several events are suggested to trigger the death pathway during seed ageing. The glutathione half-cell reduction potential $E_{GSSG/2GSH}$ increases after ROS accumulation and shifts to a zone between -180 and -160 mV which overlaps with values for human cells undergoing PCD (Schafer & Buettner 2001). Changes in the intracellular 'redox' environment could be a part of signalling cascade which is either involved in PCD stimulation at the beginning or in metacaspase activation resulting in DNA degradation (Kranner *et al.* 2006). Furthermore, depletion of the ATP pool during ageing is known to induce PCD when it is coupled with calcium influx (Lam *et al.* 2001, Kibinza *et al.* 2006). Mitochondria integrate this signal and initiate a release of cytochrome c which is considered to be a signal for the cytoplasm to execute a PCD (Jones 2000). Additional investigations of gene expression patterns revealed a hormonal control by jasmonic acid and ethylene via ethylene-responsive element (ERE) binding protein transcription factors (Sreenivasulu *et al.* 2006). The death pathway of barley aleurone cells indicates a strong influence of gibberellic acid which results in a decline of ROS scavenging enzymes and inhibits the ability to metabolise ROS (Fath *et al.* 2001).

Unfortunately, most studies about seed ageing and PCD are carried out via controlled deterioration which increases the moisture level of seed. In this state both PCD and non-programmed cell death are likely to operate. Due to the limited molecular mobility of the cytoplasm in the glassy state and lack of water for biochemical processes direct oxidative damage close to their production site is more probable (Kranner *et al.* 2010b).

1.1.5 Dormancy

Programmed cell death, lipid peroxidation and DNA degradation are some of many detrimental effects caused by reactive oxygen species. Since the control of signalling pathways and dormancy release are known to be beneficial for seed germination and seedling growth, the role of ROS has been reconsidered (El-Maarouf-Bouteau & Bailly 2008).

In this regard seed dormancy is an important property which reduces competition between individuals of the same species, prevents germination out of season and may ensure the survival in case of natural disasters (Finkelstein *et al.* 2008). It can be thought as a block to complete germination of an intact viable seed under favourable conditions (Bewley 1997, Finch-Savage & Leubner-Metzger 2006). Generally, dormancy is distinguished into primary dormancy that is acquired during seed maturation and secondary dormancy that seeds enter when they are imbibed after shedding and exposed to unfavourable temperature conditions or light and nitrate deficiency (Finkelstein *et al.* 2008). Secondary dormancy has been further classified into five groups: physiological (PD), morphological (MD), morphophysiological (MPD), physical (PY) and combinational dormancy (PY+PD) (Baskin & Baskin 2004, Finch-Savage & Leubner-Metzger 2006).

Physiological dormancy (PD) is the major form of dormancy and can be divided into deep, intermediate and non-deep PD. The latter is the most common which is releasable by scarification, after-ripening in dry storage, cold or warm stratification. Deep PD reacts only partly by applying these treatments or produces abnormal seedlings (Baskin & Baskin 2004). The morphological dormancy (MD) appears in seeds with small embryos compared to the entire seed size and is proposed to be the ancestral dormancy type among gymnosperms and angiosperms (Forbis *et al.* 2002). Species underlying MD are characterized by the embryo to seed ratio (E:S ratio) that describes the relative size of the embryo within a seed. Species which seeds have a low E:S ratio, consequently a tiny embryo and a high amount of storage tissue, have usually long germination times (Finch-Savage & Leubner-Metzger 2006, Linkies *et al.* 2010). The MD simply delays germination due to the time that the embryo needs to grow inside the seed (Linkies *et al.* 2010). This is also specific for morphophysiological dormancy (MPD) except the fact that a dormancy-breaking treatment is also required. In contrary physical dormancy (PY) is caused by water-impermeable layers in seeds and fruit coats which control water movement. A cutting of seed coat releases PY which is not the case when it is combined with PD (PY+PD) (Finch-Savage & Leubner-Metzger 2006).

Up to now the mechanisms behind this phenomenon are poorly understood but the influence of the genetic background of species, the environmental conditions and the balance of ABA and GA are known features (El-Maarouf-Bouteau & Bailly 2008).

Oracz *et al.* (2007) and El-Maarouf-Bouteau *et al.* (2007) demonstrated as one of the first that there is also a clear relationship between seed dormancy alleviation and accumulation of ROS and peroxidation products in sunflower (*Helianthus annuus* L.) (Fig. 4). Seeds undergoing after-ripening, which defines dormancy alleviation during a period of low-moisture contents (Leubner-Metzger 2005), are proposed to use ROS for signalling roles in gene expression towards germination. The ROS accumulation occurred concomitantly with carbonylation of specific embryo proteins suggesting a preparation of storage protein mobilisation (Oracz *et al.* 2007). Additionally, H_2O_2 seems to have a role in cellular response to ABA by regulating ion movements through Ca^{2+}-permeable channels in the plasma membrane (Pei *et al.* 2000) and are also assumed to inactivate ABI1 and ABI2 type 2C protein phosphatases which both are necessary for ABA signalling (Meinhard & Grill 2001, Meinhard *et al.* 2002).

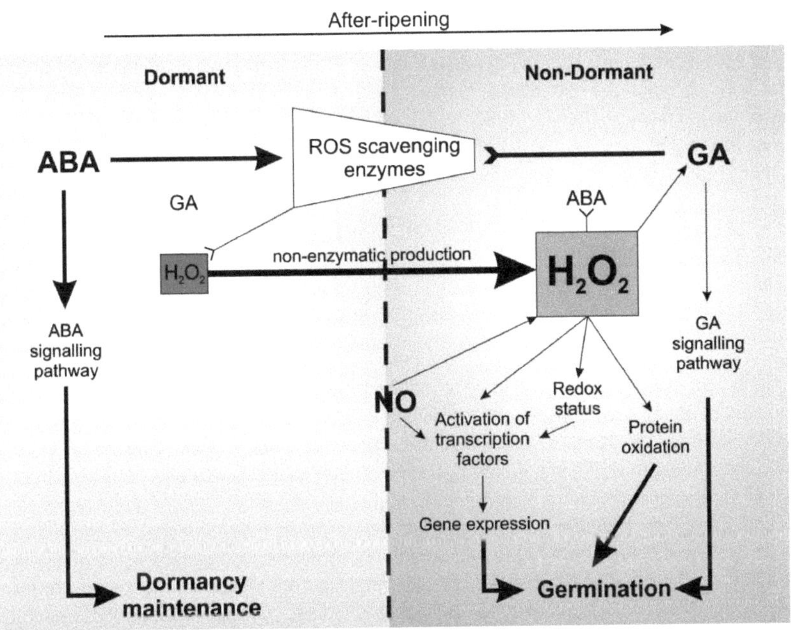

Fig. 4: **Central role of ROS in seed dormancy release and germination**. Hypothetical model proposed by El-Maarouf-Bouteau & Bailly (2008) shows that a high amount of ABA induce active signalling pathways to maintain dormancy. A depletion of ABA concentration is associated with a higher ROS level, presumably H_2O_2, which modifies redox status and protein carbonylation and is assumed to interfere with GA. ABA, abscisic acid; GA, gibberellic acid; H_2O_2, hydrogen peroxide; NO, nitrate; ROS, reactive oxygen species.

ABA is recognized as being the most important player in dormancy maintenance. It is produced during seed development and ABA *de novo* synthesis has been reported during imbibition of dormant seeds (Hilhorst 1995, Koornneef *et al.* 2002). The intrinsic balance of ABA and GA reflects the dormancy status of seed. Maintenance is achieved at high ABA/GA ratios. Dormancy release involves increase of GA biosynthesis and ABA degradation resulting in low ABA/GA ratios (El-Maarouf-Bouteau & Bailly 2008). Thereby, GA has been described as an internal regulator involved in the induction of germination (Bewley 1997). It induces hydrolytic enzymes to weaken barrier tissues such as endosperm or seed coat, enhances mobilisation of seed storage reserves and stimulates expansion of the embryo (Finkelstein *et al.* 2008). Also, ethylene is proposed to suppress dormancy by ABA inhibiting (Kucera *et al.* 2005). Further hormones as auxin or brassinosteroids seemed to integrate in this extensive hormonal cross-talk and shift conditions among their respective signalling pathways to germination (Finkelstein *et al.* 2002).

Summarising, seed dormancy is a complex phenomenon which is mainly subjected to environmental conditions that result in signal transductions including hormonal reactions by changing ABA/GA balance (El-Maarouf-Bouteau & Bailly 2008).

1.1.6 Seed coat

The seed coat is, as a part of the seed survival strategy, involved in physical dormancy (PD) and maintains, through an impermeable barrier, the environment around the embryo which is conducive for quiescence (Bewley 1997, Moise *et al.* 2005). In general, the seed coat is a multifunctional organ that plays an important role for seed development and is the primary defence organ against unfavourable environmental conditions (Debeaujon *et al.* 2000).

A hard seed coat is able to protect the seed from mechanical stress, microorganism invasion and temperature and humidity fluctuations during storage (Mohamed-Yasseen *et al.* 1994). Daws *et al.* (2007) and Shen-Miller (1995, 2002) could even prove that this water-impermeable layers ensure long-term survival and are positively correlated with seed coat colour (Debeaujon *et al.* 2000).

Pigmentation of seed coats is generally caused by phenolic compounds which are accumulated dependent on the species. Isoflavons are for example common in legumes and proanthocyanidins and flavonol glycosides are present in *Arabidopsis*. The latter ones become cross-linked to wall components by oxidation during seed maturation and construct thicker cell layers which enable a higher mechanical strain and reduce permeability to water, gases and hormones (Pourcel *et al.* 2007). As antioxidants they also control ROS and metabolic processes during after-ripening and provide protection against microorganisms and UV light. The degree of pigmentation is dependent on the photoperiod experienced by the mother plant. Strongly pigmented seeds are able to filter the light reaching the embryo (Winkel-Shirley 2002).

Genetic variation in structure and/or pigmentation of the seed coats or surrounding layers, leads to altered dormancy and seed longevity in many species (Finkelstein *et al.* 2008). The red seeds of charlock (*Sinapis arvensis* L.) show a reduced dormancy compared with black seeds (Debeaujon *et al.* 2000, Rajjou & Debeaujon 2008). White legume seeds imbibe faster than coloured seeds but suffer also greater imbibitional damage which affects viability (Powell 1989). In wheat (*Triticum aestivum* L.), high resistance to pre-harvest sprouting is associated with a red seed coat colour, whereas the lines with white seed coats are susceptible (Groos *et al.* 2002). *Arabidopsis* mutants affected by a structural aberration showed reduced dormancy, germination and storability (Debeaujon *et al.* 2000).

1.2 Genebanks and plant genetic resources

1.2.1 Genetic bottleneck of crops

Since the mid-cretaceous angiosperms and their orthodox seed have spread around the world and integrated themselves to their surrounded environment. A huge variety of protection and defence mechanisms evolved and ensured the life for the next generation. With the beginning of agriculture about 10,000 years ago, Neolithic farmers exerted a strong selection pressure on these wild progenitors of our today's crops and caused a rapid and radical change in plant species (Tanksley & McCouch 1997). In the area of South Eastern Turkey/ Northern Syria, also called the Fertile Crescent, seven founder crops (einkorn wheat (*Triticum monococcum* L.), emmer wheat (*T. dicoccum* Schrank), barley, lentil (*Lens culinaris* Medik.), pea (*Pisum sativum* L.), bitter vetch (*Vicia ervilia* Willd.), and chickpea (*Cicer arietinum* L.)) were domesticated (Lev-Yadun *et al.* 2000) according to certain traits such as non-shattering of seeds, compact growth habit or loss of dormancy (Harlan 1975). These were processes that gave rise to landraces over thousands of years (Damania 2008). Continuous domestication reduced the genetic variation but the decline appeared dramatically with modern plant breeding (Fig. 5) (Tanksley & McCouch 1997).

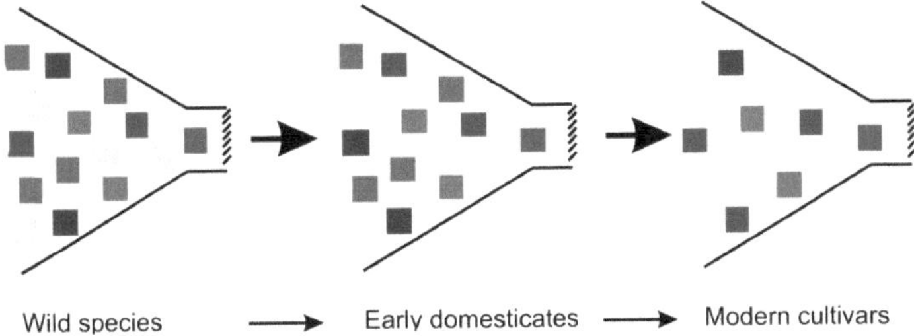

Fig. 5: Genetic bottleneck of crop plants caused by domestication and through modern plant-breeding practice. Boxes represent allelic variations of genes originally found in the wild, but gradually lost through domestication and breeding (Tanksley & McCouch 1997).

In the mid of the 19th century professional plant breeding began and was intensified with the rediscovery of Mendel's work on genetic inheritance. When humans faced the problems of World Wars, industrialisation and a rise in populations especially in developing countries, excessive funding of food aid and research programmes were initiated and led to the Green Revolution. Within decades the introduced crop germplasm increased yields but caused also an elimination of cultivars that were the indigenous products of several hundred years of development and selection (Damania 2008).

Ironically, it is the plant breeding itself which threatens its genetic basis and exposes new cultivars to increased risks of yield losses caused by pathogens, insects or environmental stresses (Tanksley & McCouch 1997). Between 1846 and 1851 Ireland had to experience the awful consequences. The usage of only a couple of potato cultivars with a highly similar genetic background cut the potato production by half due to the late blight disease which was caused by the fungus *Phytophtora infestans*. About 1 million Irish died of starvation or associated diseases and another 1.5 million left the country (Miller & Wagner 1994).

1.2.2 The heritage of plant hunters

At the beginning of the 20th century, it was Nicolai I. Vavilov who first called attention to the potential of crop relatives as a source of genes for improving agriculture (Tanksley & McCouch 1997). Due to basic research in plant genetic resources, his theories regarding the origin of crops, their evolution, their variability in space and time (Vavilov 1926, 2009) and more than 100 plant collecting missions, Vavilov is considered to be the father of crop plant exploration and collection (Hawkes 1983). His American colleague Frank N. Meyer was not a man of theories but introduced about 2,500 plant varieties from the whole world to the United States Department of Agriculture at the same time. Many others followed to gain an improvement for their local breeding lines (Damania 2008).

Based on Vavilov's enthusiasm for plant collections the former Soviet Union gained an early lead in collecting and conserving plant genetic resources. The Genetic Department at the All-Union Institute of Plant Breeding (VIR) in St. Petersburg (formerly Leningrad) and the first genebank were set up by Vavilov himself and became responsible for collecting and conserving global plant diversity (Zakharov 2005). Stimulated by Vavilov's ideas, in the mid 1930's the establishment of a German 'Institute of Cultivated Plant Research' was planned by leading Scientists like Fritz von Wettstein, Erwin Baur and Hans Stubbe. The aim was to apply basic research to crop plants and related plant species by using material from the *ex situ* genebank derived from various collection missions (Wobus & Schubert 2002). Since the beginning of the 20th century the United States have also performed collection and evaluation missions on plant genetic resources. However, the first repository of crop seed was not established until 1947 (Damania 2008).

Up to 1975, eight long-term genetic resources conservation centres were set up globally. Only seven years later the number increased to 33 (Damania 2008) and jumped to over 1,750 germplasm collections in genebanks all over the world today (FAO 2010). The enormous increase of germplasm centres is especially due to the creation of the International Board for Plant Genetic Resources (IBPGR), today Bioversity International, in 1974 (Bioversity 2011). In 1992, the Convention on Biological Diversity has also started its work and adopted the Global Strategy for Plant Conservation. Priorities have been set for threatened species and regions and for the development of protocols and models to enhance the conservation options for both crops and wild species (CBD 2002). Since this time the total worldwide protected area (*in situ* conservation) has expanded to 17.5 million km². Similarly the *ex situ* collections of crops increased to 7.4 million accessions (FAO 2010).

1.2.3 Genebanks

In situ and *ex situ* approaches should be regarded as complementary methods that provide different types of protection for the species. As an insurance policy against extinction and with estimated costs as little as 1% of *in situ* conservation, the *ex situ* conservation provides a more efficient and economic tool to preserve the world's biodiversity (Li & Pritchard 2009).

The *ex situ* maintenance of germplasm can be carried out as conservation of clonal crops or recalcitrant seeds in field genebanks, as tissue cultures or through cryopreservation and the long-term storage of orthodox seeds in seedbanks (Li & Pritchard 2009). The largest crop collections are housed at the National Center for Genetic Resources Preservation in the USA, the Institute of Crop Germplasm Resources in China, the National Bureau of Plant Genetic Resources in India and the Vavilov All-Russian Scientific Research Institute. Further national banks with more than 100,000 accessions are found in Brazil, Canada, Germany, Japan and the Republic of Korea. The largest seed genebank devoted to wild species is currently the Millennium Seed Bank run by the Royal Botanical Gardens KEW in the United Kingdom (FAO 2010).

The majority of the *ex situ* stored material is maintained as seed (FAO 2010). Especially seeds of direct national interest for food and energy supply are targeted and belong to landraces, old cultivars and related wild species. Cereals are at the top of the collected species followed by food legumes, forages, vegetables and fruits (Fig. 6). For the maintenance the FAO/IPGRI standards (1994) give a detailed description for optimum germplasm conservation. Genebanks working according to these standards split an accession in a sample for the base collection for long-term storage and for the active collection which is easy accessible. The base seed samples are dried to moisture content between 3 and 7%, filled in airtight containers and kept at subzero temperatures (< 0°C). The active seed samples are handled differently between genebanks and can be even stored in paper backs under ambient conditions (Damania 2008). Currently these standards are renewed and adapted to the present needs (personal communicated by Andreas Börner).

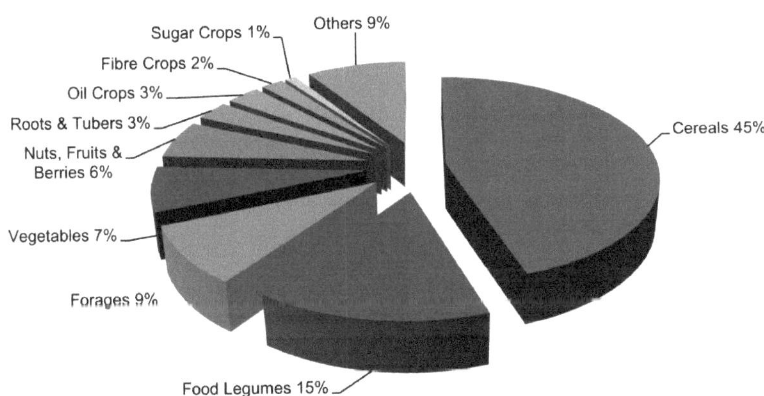

Fig. 6: Contribution of major crop groups to total *ex situ* collections. About 45% of all accessions in world's genebanks are cereals followed by food legumes and vegetables, fruits and forage crops (FAO 2010).

However, the FAO (2010) could gather little information about storage conditions in single countries. Especially developing countries have massive problems to meet the standards due to less reliable utilities and limited funding. The inestimable conservation situation in these countries was one of many reasons to set up a global germplasm conservation facility in 2008. The Svalbard International Seed Vault provides physical security because it is a backup of the world's germplasm collections and saves them in case of local catastrophes (Hopkin 2007).

16

1.2.4 Challenges and perspectives

Despite the economic superiority of *ex situ* genebanks they also face a vast majority of problems which need to be solved in the future.

The situation of germplasm collections maintained in developing countries is hard to estimate but even the viability status of material stored in high standard facilities is partly unknown (FAO 2010). Seeds will lose their viability during storage but dependent on the conditions the time is alterable (Priestley *et al.* 1985, Walters *et al.* 2004, Walters *et al.* 2005a, Nagel & Börner 2010). Additionally, genotypic differences in the longevity of species (Nagel *et al.* 2010) make it impossible to predict optimum storage periods and require frequently viability testing which is labour- and cost-intensive. New monitoring techniques such as infrared thermography demonstrated by Kranner *et al.* (2010a) could enable a faster testing and a better prediction if material needs to be regenerated. Also cryopreservation (Walters *et al.* 2004, Li & Pritchard 2009) or storage under vacuum or nitrogen (Barzali *et al.* 2005) is to be thought to extend the storage period, to reduce expenditure and the risk of loss during regeneration.

The regeneration itself is additionally seen as problematic event. The procedure is supposed to ensure the genetic integrity and to minimize genetic erosion of an accession (Börner 2006). Nevertheless, it also exposes plants to another environment than derived and that results in stress. Furthermore, only a small part of an accession can be regenerated which is associated with loss of genetic variation and genetic drift (Ramanatha Rao & Hodgkin 2002). Especially cross-pollinators seem to be susceptible to genetic shift. Studies on rye revealed variations in allele frequencies after long-term storage and several regeneration cycles (Börner 2006, Chwedorzewska *et al.* 2006), whereas self-pollinators like wheat (Börner 2006) and rice (*Oryza sativa* L.) (Tabien *et al.* 2008) showed a high degree of identity after 50 and 40 years of storage, respectively. However, there is a need to generate species specific information on regeneration techniques, population size and genetic drift (Ramanatha Rao & Hodgkin 2002).

Information about unique genetic material is also key to unlock their genetic potential. While it is believed that useful genes are maintained in genebanks the vast majority of these accessions make no contribution to modern cultivars, especially with respect to complex traits such as yield or nutritional quality (Tanksley & McCouch 1997). The traditional approach was to screen accessions for a clear defined phenotype which is controlled by one or few genes. Nevertheless, an important trait as yield is controlled by many genes and can only be evaluated directly for the presence or absence of desirable genes. Molecular maps, Quantitative Trait Locus (QTL) analysis (Tanksley & McCouch 1997), whole-genome association mapping (Rostoks *et al.* 2006), allele-mining at molecular level (Bhullar *et al.* 2009) and TILLING (Targeting Induced Local Lesions IN Genomes) techniques (Weil 2009) are among others the powerful tools for this way.

1.3 Barley (*Hordeum vulgare* L.)

Barley (*Hordeum sp.*) is with about 470,000 accessions among of the three most held species in genebanks worldwide (FAO 2010). Generally, this crop was domesticated from its wild progenitor *Hordeum spontaneum* and belongs to the seven Neolithic founder crops which were found in the core area of the Fertile Crescent (Lev-Yadun *et al.* 2000). The earliest evidence of its use comes from charred plant remains of wild barley, which were dated to be 19,000 years old (Kislev *et al.* 1992) and indicate the collection of wild barley by hunter-gatherer communities from surrounding area before domestication (Kilian *et al.* 2010). Oldest remains of cultivated barley were found in the Fertile Crescent and date back to 8,000 BC (Zohary & Hopf 2000). Since this time the domesticated two-rowed and higher yielding six-rowed varieties (Komatsuda *et al.* 2007, Pourkheirandish & Komatsuda 2007) have spread with other crops through the Mediterranean area to Europe and Africa, and eastwards through Iran and Afghanistan into India and China (Kilian *et al.* 2010). Today it is grown over a broad range of environmental conditions from 70°N in Norway to the 46°S in Chile including Tibet, Ethiopia and the Andes (Grando & Gormez Macpherson 2005).

In 2009, barley production ranks fourth in the world after maize (*Zea mays* L.), rice and wheat having a volume of 150 million tonnes (FAO 2011). Barley caryopses, henceforth termed seeds, are used as feed for animals, human food and in the malting industry. During the Roman Empire this cereal was the energy food and awarded for Gladiators called *hordearii*, barleymen, due to the main component of their training diet. Since the first century AD a shift to wheat as human food was initiated in the Mediterranean area and barley lost its main importance (Grando & Gormez Macpherson 2005, Hayes & Szucs 2006). Today it remains still substantial in some cultures around the world, particularly in Asia and Northern Africa. Furthermore, there are new perspectives because of its interesting nutritional value notably the β-glucans on lowering blood cholesterol levels and its association with increased satiety and weight loss (Baik & Ullrich 2008). From the production point of view its early maturation and high adaptability to abiotic stress including cold, drought and salinity makes it perfectly suited for cultivation from boreal to equatorial regions (Schulte *et al.* 2009).

During the last decade barley turned into an experimental model for the grass family (*Poaceae*) and its biology. Reasons for this are explained by its simpler diploid (2n=14) genome in comparison with the hexaploid wheat (Schulte *et al.* 2009) and the close relationship with wheat, rye (*Secale cereale* L.) and their wild relatives (Gaut 2002). This includes that seeds share a common architecture which is defined by an embryo attached to a starchy endosperm via a tissue (scutellum) delivering nutrients during germination (Schulte *et al.* 2009).

A broad spectrum of resources has been developed to enable the systematic analysis of barley and its genome including large numbers of mapped molecular marker, EST (expressed sequence tags) and mutant collections, DNA arrays and transformation protocols (Sreenivasulu *et al.* 2008). Further advances towards physical map construction and sequencing will be contributed by an International Barley Genome Sequencing Consortium that is funded by eight institutions (Schulte *et al.* 2009). New technological possibilities are achieved in barley genomic and genetics by Diversity Arrays Technology (DArT) (Wenzl *et al.* 2004, Akbari *et al.* 2006) and the GoldenGate Assay (Hyten *et al.* 2008) allowing a high-throughput sequencing as basis for high-density maps and association mapping which creates a detailed mapping resolution and is less time consuming (Hall *et al.* 2010).

1.4 Objectives

Against the background of climate change, overexploitation, pollution, invasive alien species and the usage of transgenic plants in agriculture, the protection of our biodiversity turned into an important target during the recent decades (Corvalan *et al.* 2005). 2010 was declared to be the International Year of Biodiversity by the United Nation who invited to safeguard the life on earth (CBD 2010). Thereby, the conservation of our plant genetic resources was assigned in parts to *ex situ* genebanks that act as an insurance policy against extinction by maintaining unique genotypes for several decades or centuries (Li & Pritchard 2009).

The majority of the *ex situ* stored material (in total 7.4 million accessions) is maintained as seed in about 1,700 genebanks worldwide (FAO 2010). However, despite the availability of improved methods of seed banking such as subzero temperatures and low moisture contents, there is a wide range of storage facilities and conditions. While some genebanks are able to meet the FAO/IPGRI (1994) demands, many are not, and consequently some collections are unconsciously degenerating (FAO 2010). Additionally seedbanks have to overcome problems which are related to the storage behaviour of seed. Previous studies gave indication that the storage period is not only determined by the species it is also influenced by the genetic background of single accessions (Nagel & Börner 2010) and thus adds to environmental effects, post-harvest and storage conditions. These factors have a strong influence on the survival of seeds and the regeneration periods which are labour-, time- and cost- intensive.

The main objective of the studies included in this thesis was to investigate the mechanisms underlying seed deterioration with respect to genetic diversity at different storage treatments ranging from cold storage with low seed moisture content (smc) to heat treatment with high smc. The attention was focused on barley that is characterised by its high degree of intra-specific diversity, its synteny to other grass genomes, the availability of modern techniques as for example expressed sequence tags (EST) markers for association mapping and the ease of accessibility to information about sequences of annotated functions and consensus maps.

The specific aims were:

- To investigate the existence of genotypic differences within species.
- To identify loci responsible for seed deterioration at different storage treatments.
- To link the genetic background with metabolic processes in seeds.
- To investigate the role of some metabolites (glutathione, tocochromanols, proteins, starch, oil) and its relation to seed deterioration.

References of Prolog on page 99 ff

First Paper

2 Longevity of seeds - intraspecific differences in the Gatersleben genebank collections

Manuela Nagel, Mian Abdur Rehman Arif, Maria Rosenhauer and Andreas Börner

Published in the proceedings of the 60th conference of the plant breeders and seed traders association in Austria (2010), 179 – 182

ISBN: 978-3-902559-37-1

Abstract

The federal *ex situ* genebank of Germany in Gatersleben is with some 150,000 accessions one of the largest of the world. About 130,000 accessions are kept as seeds. Information about optimum storage conditions, germinations and seed longevities are important for running processes. The current study is dedicated to the seed longevity of barley, wheat, rye, sorghum, flax and oilseed rape. Unfortunately, there is only a tendency for the long-life behaviour of a species because seed longevity itself is strongly influenced by genotypes.

Keywords: Genebank, Genetic diversity, Intraspecific variability, Seed storage

Permission obtained by Heinrich Grausgruber (BoKu Vienna) responsible for the proceedings

Introduction

Nowadays the trait 'seed longevity' is a primary feature for a running genebank. Information about longevity of species is important for storage periods, reproduction cycles and germination test intervals. Nevertheless, the trait itself was not considered highly relevant until the beginning of the 20th century. At that time, the Russian botanist and ecologist Nicolai I. Vavilov recognized, as one of the first, that our plant genetic resources are in particular danger and need to be conserved for the future (Maxted *et al.* 1997). These efforts were followed by the establishment of modern genebanks around the world (Linington & Pritchard 2001), which led to an increasing interest in the storage behaviour of seeds. Haferkamp *et al.* (1953), Priestley *et al.* (1985), Steiner & Ruckenbauer (1995), Walters *et al.* (2005), and Nagel & Börner (2010) investigated seed longevities of crop plants under ambient, cold and ultra-dry storage conditions. Consistently they agree that the longevity of seed is different between species and depends on the storage conditions. Depending on the conditions and the species the viability can last between months and decades. Moreover, different discoveries have demonstrated that under certain conditions seeds can survive for hundreds of years. The seeds of a date palm (*Phoenix dactylifera* L.) provide a famous example of extreme longevity. These seeds were discovered beneath a Heriodin fortress in Israel and germinated after 2000 years (Sallon *et al.* 2008). Furthermore, 200-year-old seeds could germinate in the Millennium Seed Bank at the Royal Botanic Gardens Kew. The material originated from a trip to the Cape of Good Hope in 1803. A Dutch merchant brought the seeds by ship to London where they were stored in the Tower of London and later on in the National Archives (Daws *et al.* 2007). However, the longevity of seeds has always defined for the species.

The following study will show, differences not only exist between species, but also between the genotypes within a species. The material under study belonged to the *ex situ* genebank for agricultural and horticultural crop plants in Gatersleben, which houses some 150,000 accessions, covering over 3,000 species within 890 genera. In general, orthodox seeds are conserved at -15°C, while vegetatively reproduced plant materials are maintained by in vitro culture and cryopreservation. Seed storage accounts for about 90% of the stored material (Börner 2006).

Material and methods

For investigating the intraspecific variability of seed longevity, Gatersleben genebank accessions of barley (*Hordeum vulgare* L.), wheat (*Triticum aestivum* L.), rye (*Secale cereale* L.), sorghum (*Sorghum bicolor* L.), oilseed rape (*Brassica napus* L.) and flax (*Linum usitatissimum* L.) were used. The material was multiplied in the Gatersleben estates and stored in glass jars topped with silica gel at 8±2% seed moisture content. Number of accessions, countries of origin, harvest year and storage temperature are given in Tab. 1.

For investigating the intraspecific variability seeds were tested in three replications consisting of 50 seeds each. Each replication was germinated according to the ISTA (International Seed Testing Association) rules. The germination percentage was calculated from the proportion of normal appearing seedlings (ISTA 2008).

Species	Accessions	Countries of origin	Harvest year	Storage temperature
Barley	50	14	1974	±0°C, since 2008: -15°C
Wheat	41	18	1974	±0°C, since 2008: -15°C
Rye	36	9	1982	±0°C; since 1998: -15°C
Sorghum	5	4	1978	±0°C, since 2006: -15°C
Oilseed rape	45	6	1983	-15°C
Flax	52	12	1978	-15°C

Results and Discussion

Germination after 26 up to 33 years of storage was assessed among the different crop species (Fig. 1A-F). Most crops showed high germination when germinated within 5 years post harvest, but germination of most accessions within species separated strongly after 20 years. In particular, wheat germination resulted between 0 and 87% after 34 years of storage (Fig. 1B) and barley accessions germinated between 43 and 95% after 35 years (Fig. 1A). A paired *t*-test showed significant differences between genotypes of the species after the recent germination test.

In general, as it has been documented by Roberts (1972) germination decreased over storage time in a sigmoid fashion, as visible in Fig. 1E, whereby the parameters of these curves seemed to be species specific. Especially rye accessions (Fig. 1C) had already reduced germination (between 8 and 47%) after 27 years. Similarly, most oilseed rape accessions (Fig. 1E) had germinations below 50%. Only few remained over 50%. Contrary to those species, different barley genotypes (Fig. 1A) decreased slightly in their germination and remained predominantly over 50%. Therefore, a tendency for the long-life of a seed is given for the species but, as seen, the genotypes of a species differ within.

Due to the same harvest year, the same cleaning methods and storage conditions of the genotypes of the species under investigation it can be assumed that differences in germination are genetically based. First ageing and germination tests with barley mapping populations give a hint that traits like plant height, husks just as abiotic and biotic factors during the season can influence the seed longevity (Nagel *et al.* 2009). Therefore, the viability equation as published by Ellis & Roberts (1980) can forecast a longevity tendency but the behaviour of a specific genotype depends on more factors than moisture content, storage temperature and initial viability and is unpredictable at the moment.

Conclusion

The investigations on genebank material showed that the seed longevity is not only different between species it also differs between genotypes of a species. Although barley and flax seeds show a long-living tendency and rye and oilseed rape a more short-living trend the genotypes vary in germinations. On basis of the identical harvest years and conditions as cleaning as storage conditions, genetic factors for seed longevity can be only assumed.

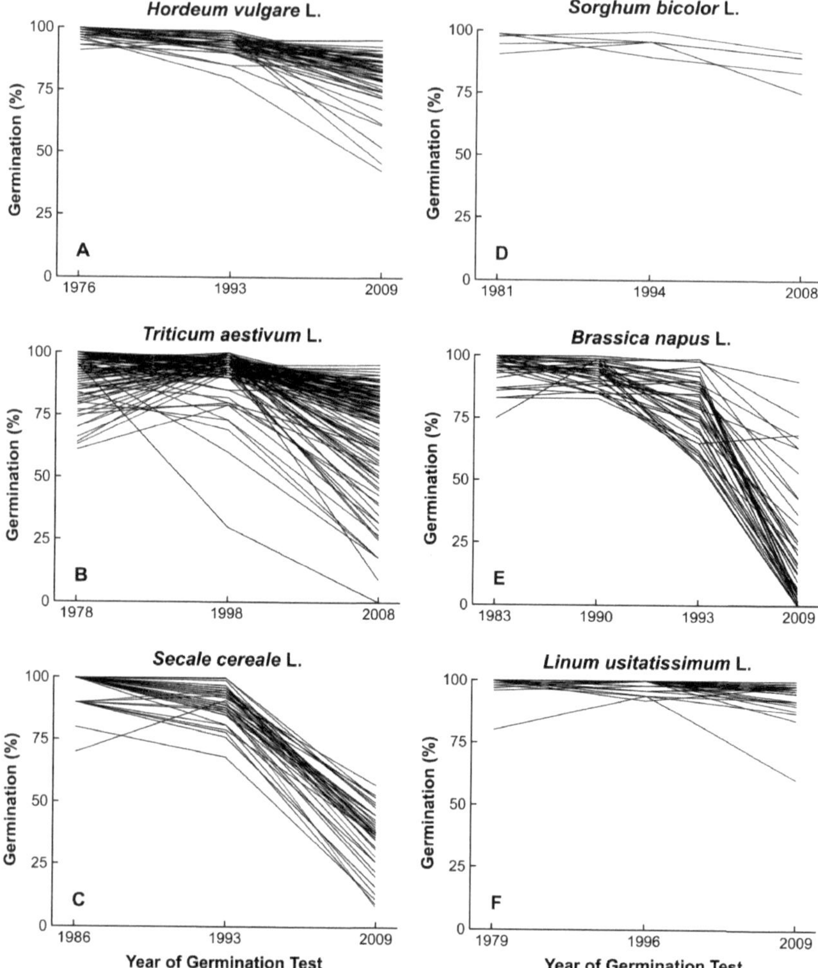

Fig. 1: Mean germinations of barley (A), wheat (B), rye (C), sorghum (D), oilseed rape (E) and flax (F) accessions in different years of testing.

Acknowledgements

The authors wish to acknowledge past and current staff of the Gatersleben genebank. Special thanks are due to Sibylle Pistrick, Stefanie Thumm and Arne Brathuhn for providing great help and scientific data.

References

Börner A. (2006) Preservation of plant genetic resources in the biotechnology era. Biotechnology Journal 1, 1393-1404.

Daws M.I., J. Davies, E. Vaes, R. van Gelder & H.W. Pritchard. (2007) Two-hundred-year seed survival of *Leucospermum* and two other woody species from the Cape Floristic region, South Africa. Seed Science Research 17, 73-79.

Ellis R.H. & E.H. Roberts. (1980) Improved equations for the prediction of seed longevity. Annals of Botany 45, 13-30.

Haferkamp M.E., L. Smith & R.A. Nilan. (1953) Studies on aged seeds,1. Relation of age of seed to germination and longevity. Agronomy Journal 45, 434-437.

ISTA. (2008) International Rules for Seed Testing International Seed Testing Association, Bassersdorf, Switzerland.

Linington S.H. & H.W. Pritchard. (2001) GENE BANKS. Encyclopedia of Biodiversity 3, 165-181.

Maxted N., V.V. Ford-Lloyd & J.G. Hawkes. (1997) Plant genetic conservation. The *in situ* approach. Chapman & Hall, London.

Nagel M. & A. Börner. (2010) The longevity of crop seeds stored under ambient conditions. Seed Science Research 20, 1-12.

Nagel M., H. Vogel, S. Landjeva, G. Buck-Sorlin, U. Lohwasser, U. Scholz & A. Börner. (2009) Seed conservation in *ex-situ* genebanks - genetic studies on longevity in barley. Euphytica 170, 5-14.

Priestley D.A., V.I. Cullinan & J. Wolfe. (1985) Differences in seed longevity at the species level. Plant Cell and Environment 8, 557-562.

Roberts E.H. (1972) Viability of seeds. Chapman & Hall, London.

Sallon S., E. Solowey, Y. Cohen, R. Korchinsky, M. Egli, I. Woodhatch, O. Simchoni & M. Kislev. (2008) Germination, genetics, and growth of an ancient date seed. Science 320, 1464-1464.

Steiner A.M. & P. Ruckenbauer. (1995) Germination of 110-year-old cereal and weed seeds, the Vienna Sample of 1877. Verification of effective ultra-dry storage at ambient temperature. Seed Science Research 5, 195-199.

Walters C., L.M. Wheeler & J.M. Grotenhuis. (2005) Longevity of seeds stored in a genebank: Species characteristics. Seed Science Research 15, 1-20.

Second Paper

3   Seed conservation in *ex situ* genebanks - genetic studies on longevity in barley

Manuela Nagel, Heike Vogel, Svetlana Landjeva, Gerhard Buck-Sorlin, Ulrike Lohwasser, Uwe Scholz & Andreas Börner

Published in Euphytica (2009) 170: 5-14

DOI: 10.1007/s10681-009-9975-7

Abstract

Recognizing the danger due to a permanent risk of loss of the genetic variability of cultivated plants and their wild relatives in response to changing environmental conditions and cultural practices, plant *ex situ* genebank collections were created since the beginning of the last century. World-wide more than 6 million accessions have been accumulated of which more than 90% are stored as seeds. Research on seed longevity was performed in barley maintained for up to 34 years in the seed store of the German *ex situ* genebank of the Leibniz Institute of Plant Genetics and Crop Plant Research in Gatersleben. A high intraspecific variation was detected in those natural aged accessions. In addition three doubled haploid barley mapping populations being artificial aged were investigated to study the inheritance of seed longevity. Quantitative trait locus (QTL) mapping was based on a transcript map. Major QTLs were identified on chromosomes 2H, 5H (two) and 7H explaining a phenotypic variation of up to 54%. A sequence homology search was performed to derive the putative function of the genes linked to the QTLs.

Keywords: *Ex situ* collections, Germplasm, Long-term seed storage, QTL mapping, Seed longevity

Introduction

World-wide existing germplasm collections contain more than 6 million accessions of which wheat represents the biggest group with about 800,000 samples followed by barley and rice comprising about 500,000 and 420,000 accessions, respectively. A list of the ten world-wide largest germplasm collections by crop is given in Tab. 1 (FAO 1998).

Tab. 1: The ten largest world-wide germplasm collections by crop (FAO 1998).

Crop	Genus	Accessions	Crop	Genus	Accessions
Wheat	*Triticum*	788,654	**Oat**	*Avena*	223,287
Barley	*Hordeum*	486,724	**Soybean**	*Glycine*	176,400
Rice	*Oryza*	420,341	**Sorghum**	*Sorghum*	168,550
Bean	*Phaseolus*	268,369	**Mustard/Rape**	*Brassica*	106,923
Maize	*Zea*	261,584	**Apple**	*Malus*	97,543

Plant *ex situ* genebank collections comprise seed genebanks, field genebanks and *in vitro* collections. Species, whose seed can be dried, without damage, down to low moisture contents, can be stored in seed banks. Field genebanks and in vitro storage are used primarily for species which are either vegetatively propagated or which have recalcitrant seeds that cannot be dried and stored for long periods. In addition, perennial species, for example certain forage species, which produce small quantities of seed, and long-lived plants (trees) are also maintained this way. It is estimated that worldwide less than 10% of genebank holdings are stored *in vivo* in the field and less than 1% are conserved *in vitro* (FAO 1998).

The German *ex situ* genebank, located at the Leibniz-Institut für Pflanzengenetik und Kulturpflanzenforschung (IPK) in Gatersleben, is one of the four largest global collections. About 150,000 accessions are maintained including cereals (65,000), legumes (28,000), vegetables (18,000), forage crops (14,000), oil crops (8,000), potatoes (6,000) and medicinal and spice plants (6,000). As on the global scale wheat (*Triticum*) and barley (*Hordeum*) are the largest groups having 28,000 and 21,000 accessions, respectively (Anonymus 2008).

Seed storage is managed in large cold chambers, maintained at 0°C or at -15°C. Seeds are kept in glass jars, covered with bags containing silica gel. The oldest material in cold storage is originated from the harvest 1974. Natural aged accessions of barley (*Hordeum vulgare* L.) being from the 1974 harvest and kept at 0°C were selected and used to study the intraspecific variability of seed longevity. In addition, three barley doubled haploid mapping populations were exploited to study the inheritance of seed longevity by performing accelerated ageing tests.

Materials and methods

Plant materials

For investigating the intraspecific variability of seed longevity 55 barley accessions comprising 45 cultivars, 6 breeding lines and 4 landraces and belonging to the subspecies/ varieties *Hordeum vulgare* L. *convar. distichon* (L.) Alef. var. *nutans* (Rode) Alef., *Hordeum vulgare* L. convar *distichon* (L.) Alef. var. *erectum* (Rode) Alef., *Hordeum vulgare* L. convar. *distichon* (L.) Alef. var. *medicum* Körn., *Hordeum vulgare* L. convar. *vulgare* var. *hybernum* Vib., *Hordeum vulgare* L. convar. *vulgare* var. *rikotense* Regel, *Hordeum vulgare* L. convar. *vulgare* var. *nigrum* (Willd.) Link, *Hordeum vulgare* L. convar. *vulgare* var. *parallelum* Körn., *Hordeum vulgare* L. convar. *intermedium* (Körn.) Mansf. var. *attergergii* Körn. were used. The material, originating from 12 countries was harvested in 1974 and stored in glass jars at $0\pm1°C$ and $8\pm2\%$ seed moisture content. Germination data were available from 1976 (initial germinability) and 1993. Together with the present investigation performed in 2008, data after 19 and 34 years of storage were available, beside the basic germination records.

For the genetic studies three different doubled haploid mapping populations were investigated. (1) The 'Steptoe' x 'Morex' (SxM) population, consisting of 150 doubled haploid (DH) lines developed by pollinating the F1 hybrid of the cv. 'Steptoe' x cv. 'Morex' cross with *H. bulbosum* (Kleinhofs *et al.* 1993). 'Steptoe' is a high yielding six-rowed feedtype barley (Muir & Nilan 1973) and 'Morex', a six-rowed cultivar used as the American malting industry standard (Rasmusson & Wilcoxson 1979). Seeds of 94 DH lines used in the present study were originated from 2006 field trials. (2) The OWB mapping population, a set of 94 spring barley DH lines developed again by the *H. bulbosum* method (Costa *et al.* 2001). Characteristics of 'OWB$_{DOM}$' and 'OWB$_{REC}$', which were selected as dominant and recessive morphological marker stocks, are described by Wolfe & Franckowiak (1991). Seeds investigated derived from field and greenhouse multiplications in 2006. (3) The winter barley population W766, resulting from a cross between the two-rowed cultivar 'Angora' and the accession 'W704/137', a two-rowed, short-stemmed, dense-eared winter barley of Japanese origin. DH lines had been derived using in vitro culture of anthers of which 100 were used in the current investigation. Further details of the mapping population are given by Buck-Sorlin (2002). Seeds were obtained from a 2005 field multiplication.

Accelerated ageing tests

The AA test exposes seeds for short periods to the two environmental variables which cause rapid seed weakening; high temperature and high relative humidity. High vigour seed lots will withstand these extreme stress conditions and deteriorate at a slower rate than low vigour ones (Hampton & TeKrony 1995).

Artificial Ageing (AA) test: For each DH line 2 replicates of 100 seeds were placed in a stainless metal cage on a rack in glass jars containing 200 ml deionised water. The ageing was performed at $43\pm0.5°C$ for 72 h.

Controlled Deterioration (CD) test: After determining the original moisture content of the seeds following the ISTA standards (ISTA 2008) it was increased to 18% by adding deionised water according to the following formula recommended by ISTA (Hampton & TeKrony 1995):

$$m_{H_2O} = \left(\frac{100 - smc_I}{100 - smc_T} - 1 \right) \cdot m_I$$

Where m_{H_2O} = added water, smc_I = initial seed moisture content (%), smc_T = target seed moisture content (%) and m_I = inital seed weight (g).

After a 2 h period of equilibration and a 22 h period of relaxation at 7°C the two replications of 100 seeds per genotype were sealed in aluminium bags and aged at 44±0.5°C for 72 h.

Germination test

For investigating the intraspecific variability of seed longevity, the 55 barley accessions were tested in three replications consisting of 50 seeds each. From the mapping populations two replications of 100 seeds each of all DH lines and originated from the field multiplications were tested as controls supplemented by one replication of 100 seeds of the OWB population originated from the greenhouse. After two different ageing methods (AA, CD), two replications of 100 seeds per method of all DH lines divided by two were analysed, i.e. four replications with 50 seeds each, per line and ageing treatment. All seeds were originated from the field multiplications mentioned above except the OWB ones used for AA method which were originated from the greenhouse.

Each replication was germinated between two layers of filter paper, formed to rolls standing on a Jacobsen apparatus (day 25 (± 2)°C; night 23 (± 2)°C). After a defined period of 7 days (ISTA 2008) the germination percentage was calculated from the proportion of normal appearing seedlings.

Statistical analysis

For testing the intraspecific variability of germinability the arithmetic means and standard deviations of the 55 different barley accessions were calculated. A paired *t*-test was applied.

Both absolute germinabilities obtained after accelerated ageing and relative germination gained by dividing the germinabilities of the single replicates and their mean by the mean germination of the corresponding controls were determined. QTL analyses were performed using absolute and relative germinabilities for each replicate and the means separately. Single marker and simple interval mapping options provided by QGENE software (Nelson 1997) were exploited. A LOD threshold of 3.0 was set to claim significance.

For OWB and SxM mapping populations transcript maps consisting of 586 and 312 expressed sequence tag (EST)-based markers, respectively, developed by Stein *et al.* (2007) are available. Markers were designated as GBR, GBM and GBS (for Gatersleben

28

barley RFLP, microsatellite and SNP). Markers in a ± 10 cM interval of the marker detected by single marker QTL analysis and having LOD values > 3 in at least one replication were considered only. Annotation of the ESTs was performed by BLASTX (Basic Local Alignment Tool) similarity search against the public non-redundant protein database, NRPEP (version from June 2008), from NCBI (National Center for Biotechnology Information). Candidate orthologs were defined as those with hits with best high scoring pair and significant E-value (Expect value) of < 1.0E-10. The sequence information of the barley ESTs are stored in the CR-EST database (The IPK Crop EST database, v1.5) (http://pgrc.ipk-gatersleben.de/cr-est).

Results

Intraspecific variability

The initial germination test showed high germination for all accessions having a mean of 99.06 ± 1.08%. The average decreased after 19 and 34 years of storage to 93.60 ± 3.26% and 76.18 ± 17.36%, respectively. There was a clear increase in variation, being highly significant in the 2008 analysis (paired *t*-test). Germination after 34 years of storage ranged between 3.36% for accession HOR 2110 and 98.67% for HOR 1320 (Fig. 1). There were no differences detectable between subspecies with respect to longevity. In fact both accessions reaching highest (98.67%) and lowest (3.36%) germinabilities belong to the same subspecies and even variety *Hordeum vulgare* L. convar. *distichon* (L.) Alef. var. *nutans* (Rode) Alef.

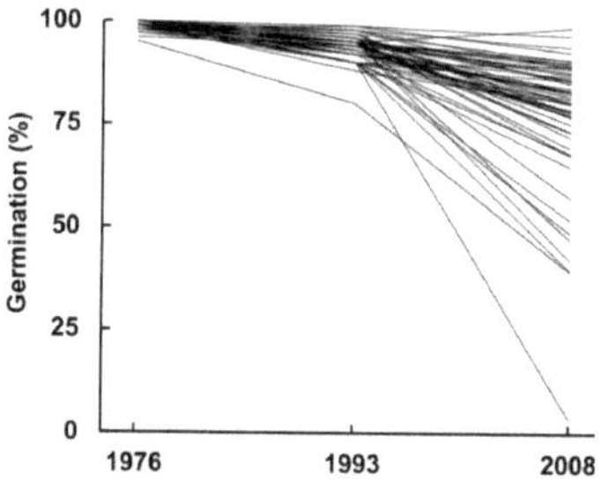

Fig. 1: Mean germination of selected barley accessions in different years of testing.

QTL mapping

Comparing the accelerated ageing QTL mapping results obtained from the data achieved using absolute germinabilities with the relative germination a high coincidence was found for both ageing methods in all three mapping populations. One example for the SxM population is given in Fig. 2. Due to this correspondence the relative germination was considered for the further assessment only.

SxM population

Applying both accelerated aging methods (AA-test, CD-test) corresponding, highly significant QTLs (LODs>14) were associated with marker GBS0892 on chromosome 5HL explaining more than 50% of the phenotypic variation (Tab. 2; Fig. 2).

Tab. 2: Single marker QTL analysis of the means of relative germination rates.

Population	Treatment	Marker	Chr.	Source allele	LOD score	R^2 value
SxM	AA	XGBS0892	5H	Steptoe	14.66	0.5397
	CD	XGBS0892	5H	Steptoe	14.34	0.5320
OWB	AA	Zeo	2H	REC	3.00	0.1452
	CD	GBS0800	5H	DOM	3.37	0.1633
		GBR1478	7H	DOM	3.19	0.1660
		GBM1047	2H	REC	2.80	0.1393
W766	AA	M38E54-320	7H	Angora	3.79	0.1633
	CD	Nud	7H	Angora	8.14	0.3177

The high longevity was contributed by the parent 'Steptoe'. No other region was discovered in that population. The locus detected by both methods was designated *QLng.ipk-5H.1*. Within a region 10 cM proximal and distal, respectively, to GBS0892 six additional markers including the gene aleurain (*Ale*) and four ESTs having LODs>3 in at least one replicate were detected. The biological functions of the candidate EST based markers (GBR0482, GBR0613) are given in Tab. 3.

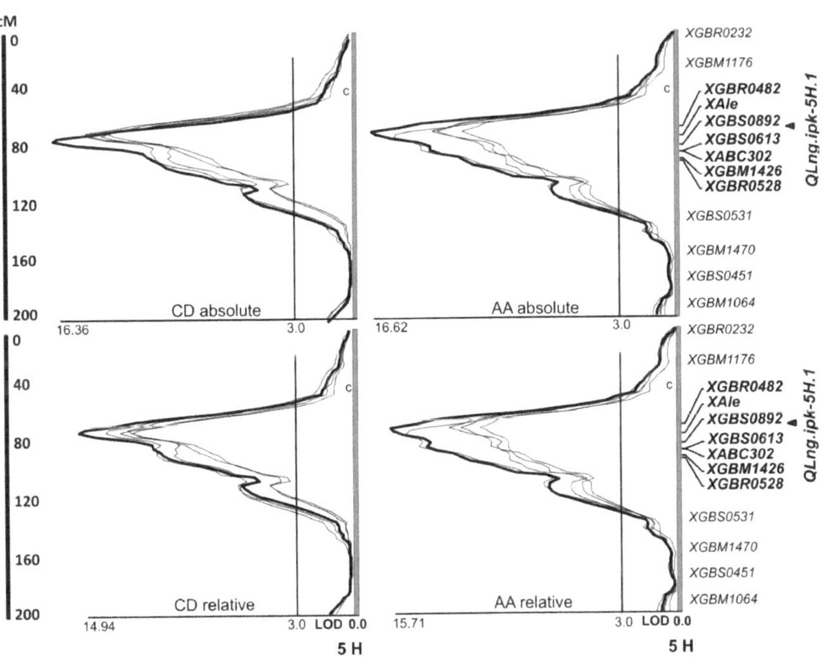

Fig. 2: QTL interval mapping results obtained for SxM population considering absolute and relative germinations in AA- and CD-tests. Skeletal map is based on data from Stein *et al.* (2007) Loci within a region of 10 cM proximal and distal to the marker detected by single marker QTL analysis and reaching a LOD>3 in at least on replicate are boxed. c = centromere

OWB population

Analysing the greenhouse originated population (AA-test) one significant QTL was detected in the distal region of chromosome 2H (Tab. 2; Fig. 3) associated with the gene locus *Zeo1* (Zeocriton 1). The same QTL, however, having a LOD value >3 in only one replicate was detected analysing the population grown in the field but applying the CD-test. Here, however, two more chromosomes were identified carrying QTLs. They are located on the distal part of the long arm of chromosome 5H connected with GBS800 and on 7HL associated with GBR1478 (Tab. 2; Fig. 3). On chromosome 7H a second peak appeared about 25 cM distal to GBR1478 reaching a significant threshold in only one replicate (Fig. 3). This region was not considered in the further analysis and discussion, because it emerged after CD-test only. The QTLs detected on chromosome 2H were contributed by parent 'OWB$_{REC}$', whereas the positive alleles on 5HL and 7HL came from 'OWB$_{DOM}$'. The QTLs explained a phenotypic variation of around 15% each. Significant (LOD>3) markers/genes within the region of 10 cM proximal and distal to the markers associated with the QTLs are indicated in Fig. 3. Candidate ESTs and their functions are given in Tab. 3. On chromosome 7H the gene determining naked caryopsis (*nud*) is in the region of interest. The QTLs detected were designated *QLng.ipk-2H*, *QLng.ipk- 5H.2* and *QLng.ipk-7H*.

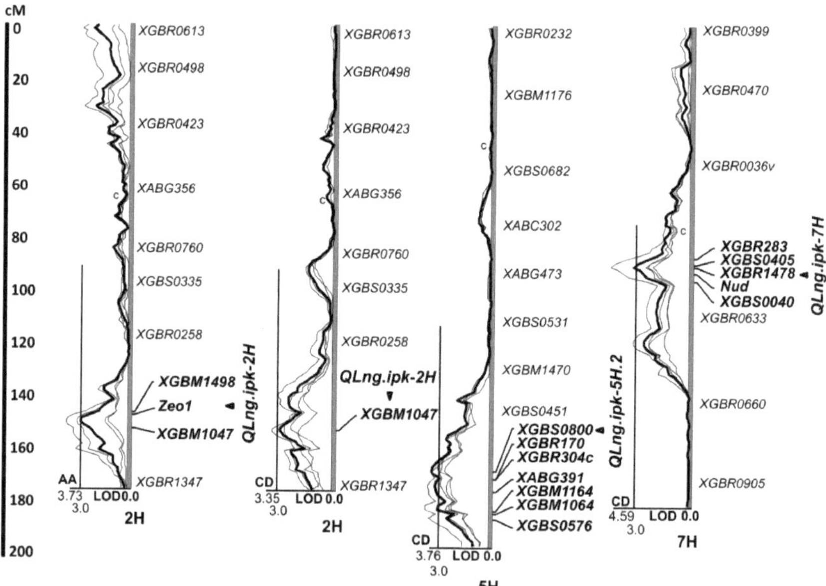

Fig. 3: QTL interval mapping results obtained for OWB population considering relative germinations in AA- and CD-tests. Skeletal map is based on data from Stein *et al.* (2007). Loci within a region of 10 cM proximal and distal to the marker detected by single marker QTL analysis and reaching a LOD>3 in at least on replicate are boxed. c = centromere

W766 population

One chromosomal region carrying QTLs for seed longevity was discovered analysing the W766 population. Corresponding QTLs were detected with both methods on the long arm of chromosome 7H close to the centromere (Fig. 4). The position was *Hordeum vulgare* L. highly comparable to that detected with CD-method in the OWB mapping population (Fig. 3). The phenotypic variation explained was 16 and 32% considering the AA-method and CD-method, respectively (Tab. 2). The higher longevity was contributed by the 'Angora' parent. Because the population was genotyped using AFLP markers no conclusions about biological functions were possible. However, the QTLs were again closely related to the gene *nud*. Because the 7H loci detected in the OWB and W766 populations are in highly comparable positions, both were designated *QLng.ipk-7H*.

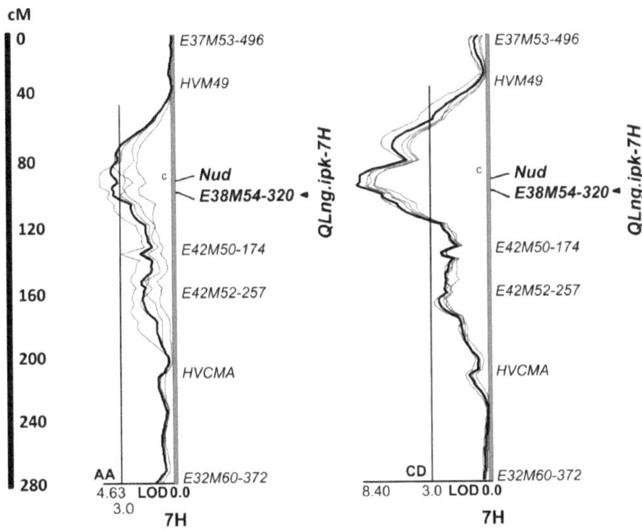

Fig. 4: QTL interval mapping results obtained for W766 population considering relative germinations in AA- and CD-tests. Skeletal map is based on data from Buck-Sorlin (2002). c = centromere

Discussion

Plant genetic resources are of particular high value for mankind. Some were collected already at the beginning of the last century and are not any longer available in the original habitats. The importance of genebank accessions is undoubted because they form the basic material for the work of plant breeders, pharmacists and ecologists. For an accurate maintenance of genebank collections consisting predominantly (90%) of seeds (Börner 2006) studies on the longevity of the accessions stored are essential.

EST-Marker	Chr.	Hit_name	Functional annotation	Organism	Score	E-value
GBM1498	2H	gi|68518815|gb|AAY98505.1|	Dehydration responsive element binding protein	*T. aestivum*	92	2E-17
		gi|68303944|gb|AAY89658.1|	DREB protein	*Glycine max*	76	9E-13
GBM1164	5H	gi|50919595|ref|XP_470158.1|	putative hydrolase	*Oryza sativa*	155	1E-36
		gi|12642902|gb|AAK00393.1|	putative epoxide hydrolase ATsEH	*Arabidopsis thaliana*	88	2E-16
GBR170	5H	gi|56682582|gb|AAW21725.1|	thaumatin-like protein TLP5	*Hordeum vulgare*	188	9E-47
		gi|2454602|gb|AAB71680.1|	Barperm1	*Hordeum vulgare*	166	5E-40
GBR304c	5H	gi|115456247|ref|NP_001051724.1|	Heat shock cognate 70 kDa protein 2, putative, expressed	*Oryza sativa*	153	6E-36
GBS0800	5H	gi|18476518|gb|AAL50205.1|	APETALA2-like protein []	*Hordeum vulgare*	409	1E-112
		gi|53830037|gb|AAU94926.1|	floral homeotic protein	*Triticum aestivum* subsp. *spelta*	394	1E-108
XGBR0482	5H	gi|115478929|ref|NP_001063058.1|	putative permease 1	*Oryza sativa*	95	2E-18
XGBS0613	5H	gi|89511843|dbj|BAE86874.1|	putative asparate aminotransferase	*Hordeum vulgare*	217	3E-55
		gi|115479507|ref|NP_001063347.1|	putative cysteine conjugate beta-lyase	*Oryza sativa*	209	9E-53
GBR1478	7H	gi|115449079|ref|NP_001048319.1|	Ethylene-responsive transcription factor 1	*Oryza sativa*	348	1E-94
		gi|6689918|gb|AAF23899.1|AF193803_1	transcription factor EREBP1	*Oryza sativa*	334	3E-90
		gi|145390028|gb|ABP65298.1|	ethylene responsive element binding protein	*Oryza sativa*	277	4E-73
GBR283	7H	gi|115475922|ref|NP_001061557.1|	putative enoyl-ACP reductase Oryza sativa	*Oryza sativa*	67	3E-10

Tab. 3: Biological functions of candidate ESTs having significant E-value (<1.0E-10).

Since the investigation of seed samples discovered in the foundation stone of the 'Nürnberger Stadttheater' in 1956 (Aufhammer & Simon 1957, Steiner *et al.* 1997) and the evaluation of the experiments initiated by the Austrian researcher Friedrich Haberland at the 'Universität für Bodenkultur' in Vienna in 1877 but rediscovered in 1967 (Ruckenbauer 1971, Steiner & Ruckenbauer 1995); it is known that cultivated crops (cereals) can keep their germinability for more than 100 years. The longevity of seeds is species specific and formulas for the prediction of seed viability have been created for many species (Ellis & Roberts 1980, 1981, Ellis 1988, Ellis *et al.* 1989). Species dependent differences in storability of seeds were also detected in studies performed at the IPK genebank considering cereals, legumes, vegetables, oil crops and herbs (Specht *et al.* 1997, Specht *et al.* 1998). However, beside this interspecific variability of seed longevity variation within a species does exist, as demonstrated in the present study on barley. The accessions investigated coming from a seed multiplication performed in the same year (1974) at the same place (experimental fields, IPK Gatersleben). Furthermore they were handled in the same way during/after harvest (threshing, cleaning) and stored under identical conditions in one and the same cold chamber in glass jars. Therefore, the differences in germinability discovered in the present study (Fig. 1) must be due to genetic variation in seed longevity. It should be mentioned here, that corresponding studies on wheat and sorghum show a very similar behaviour (unpublished data).

The consequence of these findings was to perform a genetic analysis in order to identify loci responsible for the differences in seed longevity. Comparable studies in plants are rare. One mapping study was performed in the model plant *Arabidopsis* (Bentsink *et al.* 2000).

In total four QTLs located on chromosomes 1 (two closely linked QTLs), 3 and 5 were identified. Other studies were performed in rice, where chromosomes 2, 4, 9 (two QTLs) 11 and 12 carry QTLs for seed longevity (Miura *et al.* 2002, Zeng *et al.* 2006, Shigemune *et al.* 2008). In the present study three barley mapping populations were analysed in parallel and four genomic regions associated with seed longevity QTLs were identified on chromosomes 2HL (*QLng.ipk-2H*), 5HL (*QLng.ipk- 5H.1, QLng.ipk-5H.2*) and 7HL (*QLng.ipk-7H*). Compared to QTL studies for other traits in cereals this number is rather low, whereas the phenotypic variation explained by the loci detected is high, reaching up to 54%.

In two of the mapping populations (W766 and OWB), segregating for the character hulled/naked caryopsis QTLs closely related to the *nud* gene determining the trait (Lundqvist *et al.* 1997) were identified. Therefore it is very likely, that the gene itself is involved. In both populations the hulled parent contributes to the higher longevity, i.e. naked grains are disadvantageous. The region detected on chromosome 2HL is carrying another gene, designated *Zeo1* and determining a short plant habit having very compact spikes with long awns and reduced fertility (Lundqvist *et al.* 1997). Here the high longevity is contributed by the parent 'OWB$_{REC}$', carrying the recessive wild-type allele determining 'normal' spikes. A negative effect due to the changed spike architecture may be postulated.

On chromosome 5H two regions were detected in two different mapping populations. The SxM population is segregating for the gene *Ale* (*Aleurain*) related to the QTL detected in the proximal region of chromosome 5HL. Aleurain is a barley vacuolar thiol protease. The *aleurain* gene (cDNA) was synthesized from gibberellic acid-stimulated aleurone cell mRNA (Rogers *et al.* 1985). The expression is regulated by the plant hormones gibberellic acid and abscisic acid, well known to be included in germination process. Interestingly, analysing wheat-barley single chromosome addition lines Li *et al.* (1991) localised another aleurain gene on chromosomes 7H whereas a copy of exon 3-intron 3 from the aleurain gene is present on chromosome 2H. Unfortunately no intrachromosomal mapping data are available and therefore, a comparison to the positions of the longevity loci mapped on these particular chromosomes in the present study is not possible.

The candidate ESTs identified in the longevity QTL regions have orthologs in rice, wheat, soybean and Arabidopsis or have known functions in barley itself. On chromosome 2H the marker GBM1498 has orthologs in wheat and soybean, associated with dehydration-responsive element binding (DREB) protein. Together with the ethylene-responsive element (ERE) binding factors (rice ortholog of GBR1478 on chromosome 7H) they belong to the APETALA2/ethylene-responsive element-binding protein family which play an important role in the regulation of abiotic and biotic stress responses, respectively. The expression of DREBs is activated by drought, cold or ethylene (Sun *et al.* 2008). However, mainly the osmotic stress conditions such as drought and salinity lead to induction of DREB (Huang & Liu 2006). Apart from that the ERE binding elements regulate the expression of pathogen response gene and prevent disease progression. The ethylene pathway together with the jasmonate pathway is required for regulation of defence response genes (Lorenzo *et al.* 2003).

APETALA2 (AP2) itself represents wheat and barley orthologs of the marker GBS0800 on chromosome 5H. The transcription factors of the AP2 gene family implicate a wide range of plant development roles and perform in angiosperms the establishment of the floral meristem, the specification of floral organ identity, the regulation of floral homeotic gene expression, the regulation of ovule development and the growth of floral organs (Kim *et al.* 2006). One gene of this family called 'indeterminate spikelet 1' specifies determinate fates which lead to many types of lateral organ primordial and spikelet meristems (Chuck *et al.* 1998). These may contribute to an inflorescence architecture which is important for development and physical health of seeds. In a similar way the GBR283 (chromosome 7H) ortholog, a putative enoyl ACP reductase is conducive for the ripening of fruits.

Furthermore, thaumatin-like proteins like Barperm1, which is related to the sequence of GBM1164 on chromosome 5H, is responsible for an antifungal activity and indicates a possible role in defence against leaf pathogens (Zareie *et al.* 2002). Interestingly the investigations of Tattersall *et al.* (1997) show the timing of its accumulation correlates with the inability of the fungal pathogen initiate new infections. Together with DREB and ERE binding elements it contributes to withstand environmental stress and could be one hint for different shelf-life of seeds. The effect of the orthologs of the remaining markers (Tab. 3) on seed longevity is rather speculative and therefore not discussed further.

Finally, we compared the longevity QTL regions detected here with those described in rice (Miura *et al.* 2002, Zeng *et al.* 2006) by using the colinearity information of Stein *et al.* (2007). For two of the barley QTLs co-linearity to loci in rice may be suggested. Based on the co-linearity between barley 2HL and rice Os4L the loci *QLng.IPK-2H* and *qLG-4* may represent homoeologous loci. In addition barley *QLng.IPK-5H.1* may be related to rice *qLG-9* or *qLS-9*. Ongoing studies in wheat and rye will provide more information about a possible conservation of genes for seed longevity within the *Poaceae*.

Acknowledgments

We thank Nils Stein for providing the marker data and Anita Winger, Sibylle Pistrick, Jutta Scheurenberg and Franziska Scharkowski for excellent technical assistance.

References

Anonymus. (2008) Annual report 2007. Leibniz-Institut für Pflanzengenetik und Kulturpflanzenforschung, Gatersleben. pp. 222.

Aufhammer G. & U. Simon. (1957) Die Samen landwirtschaftlicher Kulturpflanzen im Grundstein des ehemaligen Nürnberger Stadttheaters und ihre Keimfähigkeit. Zeitschrift für Acker- und Pflanzenbau 103, 455-472.

Bentsink L., C. Alonso-Blanco, D. Vreugdenhil, K. Tesnier, S.P.C. Groot & M. Koornneef. (2000) Genetic analysis of seed-soluble oligosaccharides in relation to seed storability of *Arabidopsis*. Plant Physiology 124, 1595-1604.

Börner A. (2006) Preservation of plant genetic resources in the biotechnology era. Biotechnology Journal 1, 1393-1404.

Buck-Sorlin G.H. (2002) The search for QTL in barley (*Hordeum vulgare* L.) using a new mapping population. Cellular & Molecular Biology Letters 7, 523-535.

Chuck G., R.B. Meeley & S. Hake. (1998) The control of maize spikelet meristem fate by the APETALA2-like gene indeterminate spikelet1. Genes & Development 12, 1145-1154.

Costa J.M., A. Corey, P.M. Hayes, C. Jobet, A. Kleinhofs, A. Kopisch-Obusch, S.F. Kramer, D. Kudrna, M. Li, O. Riera-Lizarazu, K. Sato, P. Szucs, T. Toojinda, M.I. Vales & R.I. Wolfe. (2001) Molecular mapping of the Oregon Wolfe Barleys: A phenotypically polymorphic doubled-haploid population. Theoretical and Applied Genetics 103, 415-424.

Ellis R.H. (1988) The viability equation, seed viability nomographs, and practical advice on seed storage. Seed Science and Technology 16, 29-50.

Ellis R.H. & E.H. Roberts. (1980) Improved equations for the prediction of seed longevity. Annals of Botany 45, 13-30.

Ellis R.H. & E.H. Roberts. (1981) The quantification of ageing and survival in orthodox seeds. Seed Science and Technology 9, 373-409.

Ellis R.H., T.D. Hong & E.H. Roberts. (1989) A Comparison of the low-moisture-content limit to the logarithmic relation between seed moisture and longevity in twelve species. Annals of Botany 63, 601-611.

FAO. (1998) The state of the world´s plant genetic resources for food and agriculture. Food and Agriculture Organization of the United Nations, Rome. pp. 510.

Hampton J.G. & D.M. TeKrony. (1995) Handbook of vigour test methods, International Seed Testing Association, Zürich.

Huang B. & J.-Y. Liu. (2006) A cotton dehydration responsive element binding protein functions as a transcriptional repressor of DRE-mediated gene expression. Biochemical and Biophysical Research Communications 343, 1023-1031.

ISTA. (2008) International Rules for Seed Testing International Seed Testing Association, Bassersdorf, Switzerland.

Kim S., P.S. Soltis, K. Wall & D.E. Soltis. (2006) Phylogeny and domain evolution in the APETALA2-like gene family. Molecular Biology and Evolution 23, 107-120.

Kleinhofs A., A. Kilian, M.A. Saghai Maroof, R.M. Biyashev, P. Hayes, F.Q. Chen, N. Lapitan, A. Fenwick, T.K. Blake, V. Kanazin, E. Ananiev, L. Dahleen, D. Kudrna, J. Bollinger, S.J. Knapp, B. Liu, M. Sorrells, M. Heun, J.D. Franckowiak, D. Hoffman, R. Skadsen & B.J. Steffenson. (1993) A molecular, isozyme and morphological map of the barley (*Hordeum vulgare*) genome. Theoretical and Applied Genetics 86, 705-712.

Li X.-Y., S.W. Rogers & J.C. Rogers. (1991) A copy of exon 3-intron 3 from the barley *aleurain* gene is present on chromosome 2. Plant Molecular Biology 17, 509-512.

Lorenzo O., R. Piqueras, J.J. Sánchez-Serrano & R. Solano. (2003) ETHYLENE RESPONSE FACTOR1 integrates signals from ethylene and jasmonate pathways in plant defense. The Plant Cell Online 15, 165-178.

Lundqvist U., D. Franckowiak & T. Konishi. (1997) New and revised descriptions of barley genes. Barley Genetic Newsletter 26, 22-516.

Miura K., S.Y. Lin, M. Yano & T. Nagamine. (2002) Mapping quantitative trait loci controlling seed longevity in rice (*Oryza sativa* L.). Theoretical and Applied Genetics 104, 981-986.

Muir C.E. & R.A. Nilan. (1973) Registration of 'Steptoe' barley. Crop Science 13, 770.

Nelson J.C. (1997) QGene: software for marker-based genomic analysis and breeding. Molecular Breeding 3, 239-245.

Rasmusson D.C. & R.D. Wilcoxson. (1979) Registration of 'Morex' barley. Crop Science 19, 293.

Rogers J.C., D. Dean & G.R. Heck. (1985) Aleurain: a barley thiol protease closely related to mammalian cathepsin H. Proceedings of the National Academy of Sciences 82, 6512-6516.

Ruckenbauer P. (1971) Keimfähiger Winterweizen aus dem Jahre 1877. Die Bodenkultur 22, 372-386.

Shigemune A., K. Miura, H. Sasahara, A. Goto & T. Yoshida. (2008) Role of maternal tissues in *qLG-9* control of seed longevity in rice (*Oryza sativa* L.). Breeding Science 58, 1-5.

Specht C.E., U. Freytag, K. Hammer & A. Börner. (1998) Survey of seed germinability after long-term storage in the Gatersleben genebank (part 2). Plant Genetic Resources Newsletter 115, 39-43.

Specht C.E., J.E.R. Keller, U. Freytag, K. Hammer & A. Börner. (1997) Survey of seed germinability after long-term storage in the Gatersleben genebank. Plant Genetic Resources Newsletter 111, 64-68.

Stein N., M. Prasad, U. Scholz, T. Thiel, H.N. Zhang, M. Wolf, R. Kota, R.K. Varshney, D. Perovic, I. Grosse & A. Graner. (2007) A 1,000-loci transcript map of the barley genome: New anchoring points for integrative grass genomics. Theoretical and Applied Genetics 114, 823-839.

Steiner A.M. & P. Ruckenbauer. (1995) Germination of 110-year-old cereal and weed seeds, the Vienna Sample of 1877. Verification of effective ultra-dry storage at ambient temperature. Seed Science Research 5, 195-199.

Steiner A.M., P. Ruckenbauer & E. Goecke. (1997) Maintenance in genebanks, a case study: Contaminations observed in the Nürnberg oats of 1831. Genetic Resources and Crop Evolution 44, 533-538.

Sun S., J.-P. Yu, F. Chen, T.-J. Zhao, X.-H. Fang, Y.-Q. Li & S.-F. Sui. (2008) TINY, a dehydration-responsive element (DRE)-binding protein-like transcription factor connecting the DRE- and Ethylene-responsive element-mediated signaling pathways in *Arabidopsis*. Journal of Biological Chemistry 283, 6261-6271.

Tattersall D.B., R. van Heeswijck & P.B. Hoj. (1997) Identification and characterization of a fruit-Specific, thaumatin-like protein that accumulates at very high levels in conjunction with the onset of sugar accumulation and berry softening in grapes. Plant Physiology 114, 759-769.

Wolfe R.I. & J.D. Franckowiak. (1991) Multiple dominant and recessive genetic marker stocks in spring barley. Barley Genetics Newsletter 20, 117-121.

Zareie R., D.L. Melanson & P.J. Murphy. (2002) Isolation of fungal cell wall degrading proteins from barley (*Hordeum vulgare* L.) leaves infected with *Rhynchosporium secalis*. Molecular Plant-Microbe Interactions 15, 1031-1039.

Zeng D.L., L.B. Guo, Y.B. Xu, K. Yasukumi, L.H. Zhu & Q. Qian. (2006) QTL analysis of seed storability in rice. Plant Breeding 125, 57-60.

Third Paper

4   Genome-wide association study and biochemical indicators of seed longevity in barley

Manuela Nagel[1], Ilse Kranner[2], Kerstin Neumann[1], Hardy Rolletschek[1], Charlotte Seal[2], Louise Colville[2], Beatrice Fernández-Marín[3], Andreas Börner[1]

[1] Leibniz Institute of Plant Genetics and Crop Plant Research, Gatersleben, Germany,
[2] Seed Conservation Department, Royal Botanic Gardens, Kew, Wakehurst Place, UK,
[3] Department of Plant Biology and Ecology; University of Basque Country; Bilbao, Spain,

Manuscript is intended for submission to Plant Journal

Summary

Against the background of 7.4 million accessions stored in genebanks, long-term survival of stored seeds is an important trait. This study intended to combine genetic and biochemical approaches to gain a broad overview over factors influencing seed longevity in barley (*Hordeum vulgare* L.). On the basis of a genome-wide association study 105 marker-trait associations (MTAs) on 32 loci were detected after artificial ageing confirming genotypic effects during long-term survival. Putative functions of MTAs and closely linked QTLs revealed predominantly biotic and abiotic stress affect seed longevity. To address aspects of abiotic, including oxidative stress, the major antioxidant glutathione (GSH) and its half-cell reduction potential ($E_{GSSG/2GSH}$) were analysed in a set of seeds that had lost viability during genebank storage (5 to 13 years) and a set of artificially aged [44°C, 13% and 18% seed moisture content (MC), up to 41days] seeds. A general depletion of the glutathione pool [GSH plus glutathione disulphide (GSSG)] and a shift towards more oxidising conditions occurred with decreasing viability. Highly significant correlations of viability with $E_{GSSG/2GSH}$ but also %GSSG were found. Plotting both parameters against each other, clusters of ageing treatments were formed presumably caused by an acidifying effect of higher seed MC levels during artificial ageing indicative of different processes occurring when different ageing conditions were used. Tocochromanols were predominately represented as tocotrienols and did not show any relationship with seed deterioration. Viability loss and storage was not correlated with changes in major storage compounds, starch, proteins and oil. In summary, it appears that barley seed ageing was affected by genetic factors, by environmental factors during seed development, by the MC during storage and by the type of ageing treatment.

Keywords: Genebank, Genome-wide association mapping, Genotypic differences, Glutathione, Half-cell production potential, Tocochromanols, Seed ageing, Storage compounds

Introduction

Barley (*Hordeum vulgare* L.) is among the four most-important crops globally and has an annual production of 150 million tonnes (FAO 2011). Barley was the first domesticated crop and is popular for breeding and science due to a high degree of natural variation (Kuckuck 1929), hermaphroditic character and synteny with other grass genomes (Gale & Devos 1998).

The seed longevity of barley was first investigated in 1877 by Haberlandt (Steiner & Ruckenbauer 1995) who stored ultra-dry seeds (3.12% moisture content (MC) at 10 to 15°C), of which 90% germinated after 110 years. At ambient conditions (~ 20°C, 50% relative humidity (RH)) barley seeds have a projected half-life of 7.2 to 9.2 years (Priestley *et al.* 1985, Nagel & Börner 2010) whilst at subzero temperatures (-18 °C and 4 - 8% MC) this can rise to 84 years (Walters *et al.* 2005b).

In general, the longevity of orthodox seed is extended by lowering seed MC and storage temperature (Harrington 1963, Roberts 1973). For optimum storage, seed MC should be decreased to 0.10 g $H_2O \cdot$(g dry mass – g lipid)$^{-1}$ (assuming $\approx$ 50% RH at $\leq$ 25°C), where the aqueous domain of the seed becomes glassy and longevity is extended (Walters 1998, Walters *et al.* 2005a). Metabolic processes are unlikely in the glassy state although may continue at very slow rates (Kranner *et al.* 2010a). Seed deterioration is associated with damage to macromolecules such as DNA, lipids and proteins. The main cause of damage are reactive oxygen species (ROS) (Bailly 2004, Halliwell 2006, El-Maarouf-Bouteau & Bailly 2008, De Gara *et al.* 2010) and non-enzymatic mechanisms such as Amadori and Maillard reactions (Sun & Leopold 1995) and lipid peroxidation (McDonald 1999).

Proteins are major target of radicals or other oxidants (Davies *et al.* 1999) resulting in permanently loss of function or a further degradation (Colville & Kranner 2010). Radicals are scavenged by antioxidants such as glutathione (GSH), tocochromanols and ascorbic acid. Under persisting stress, the antioxidant capacity decreases, resulting in a rise in ROS production and a shift in the antioxidant redox state towards more oxidising conditions (Schafer & Buettner 2001). In agreement with this concept the glutathione half-cell reduction potential ($E_{GSSG/2GSH}$) increases to more oxidising values during viability loss and is assumed to trigger further signalling cascades (Kranner *et al.* 2006).

In moist seeds, active repair mechanisms and enzymatic processes are operative (Sen & Osborne 1974, Bray & Dasgupta 1976, Sen & Osborne 1977) but consequently produce additional ROS (Schopfer *et al.* 2001, Kranner *et al.* 2010c). This increase in ROS may be perceived by membrane-bound receptor proteins, such as His kinase ATHK1 (Xiong *et al.* 2002), and results in an up-regulation of genes such as those linked to metabolism, stress response and reserve catabolism (Pestsova *et al.* 2008). Furthermore, thiol-disulphide conversions of proteins are reversed resulting in a reduction of glutathione disulphide (GSSG) and protein-bound glutathione (PSSG) to GSH and thiolated proteins, respectively (De Gara *et al.* 2003, Rhazi *et al.* 2003). A general activation of antioxidant system including up-regulation of gene expression for antioxidant synthesis is seen in wheat, pine and cress seeds during early germination (De Gara *et al.* 1997, Müller *et al.* 2010, Tommasi *et al.* 2001).

Pre-storage conditions and genotypic effects also play crucial roles in influencing longevity characteristics. Genetic studies in seed longevity have found a general genotypic difference in the ability to survive storage, identifying a range of loci linked to stress response and germination in rice and *Arabidopsis* (Miura *et al.* 2002, Clerkx *et al.* 2004, Zeng *et al.* 2006). Further markers of quantitative trait loci (QTL) detected after artificial ageing of barley were associated with important roles during plant development (Nagel *et al.* 2009).

It is not yet clear if seed deterioration and death can be attributed to one cause or a plethora of mechanisms. In addition, seeds in the glassy state with limited molecular mobility may die from different causes than seeds with higher MCs in which enzymatic activity is re-instated (Kranner *et al.* 2010a). Many studies use artificial ageing methods which reduce germination after high moisture and temperature treatment. These rapid ageing methods mimic seed behaviour in storage (Delouche & Baskin 1973, Rajjou *et al.* 2008) but comparisons to longer-term stored seed are rarely available.

The aims of this paper were to investigate how the genetic background in conjunction with the environmental conditions during the growth of the mother plant and seed storage conditions such as temperature and seed MC, affect seed longevity. We present a genome-wide association study for 175 dryland barley genotypes from 30 countries on four continents, identifying marker-trait associations (MTAs) with putative roles in longevity in response to two different artificial ageing regimes, using seeds grown in two environments. Seed longevity of another 50 selected genotypes was then tested after 34 years storage at 0°C, confirming that genotypic differences significantly affect germinability. Several MTAs indicated that protection from biotic/ oxidative stress is potentially linked to seed longevity. To address the aspect of oxidative stress, a detailed study of the effect of storage conditions and genotype on the redox state of GSH/GSSG was conducted with the seeds of 26 genotypes stored at 0 or 20°C. Of these 26 genotypes, six were artificially aged at two different MCs to compare long-term storage under seed bank conditions with artificial ageing. Moreover, the composition of the main seed storage reserves, starch, oil and protein, and the response of tocochromanols were investigated in one genotype.

Material and Methods

Association mapping

A subset of the "ICARDA population" (Varshney *et al.* 2010), obtained from the International Centre for Agricultural Research in the Dry Areas (ICARDA), Syrian Arab Republic, was used for artificial seed ageing experiments. The subset contained 122 landraces and 53 cultivars (*Hordeum vulgare* subsp. *vulgare* L.) from four continents (Africa, Asia, Australia, Europe), representing 30 countries. Seeds were multiplied in two separate experimental fields at the IPK Gatersleben estate, Germany, in 2008 and caryopses, henceforth termed seeds, were used for two artificial ageing procedures in 2009 and 2010.

For the "accelerated ageing" (AA) treatment two replicates of 100 seeds per genotype were placed above 200 ml of deionised water at $43\pm0.5°C$ for 72 h in sealed glass jars. For the "controlled deterioration" (CD) four replicates of 50 seeds were equilibrated to 47% RH above non-saturated (8.7 M) lithium chloride (LiCl) (Hay *et al.* 2008). CD was initiated by increasing the temperature to $45°C$ for 15 days, using a 7.1 M LiCl solution (60% RH) to maintain a stable seed MC. Before (I, initial) and after AA and CD, seeds were germinated on moist filter paper (four replicates of 50 seeds) under a day-night cycle (14 h day at $25\pm2°C$ and 10 h night at $23\pm2°C$). Percentage germination according to ISTA (2008) was defined as vigour, whilst total germination (TG) was assessed as radicle emergence of a minimum of 2mm and recorded as a percentage after seven days.

Genetic Diversity Array Technology (DArT) markers and their chromosomal positions were obtained from Triticarte Pty. Ltd, Canberra, Australia, and comprised 703 markers with average distances of 1.6 cM. Due to a minor allele frequency (MAF) of 16.6% (f/0.05), all markers were used.

To determine the structure of the ICARDA population, a subset of the genotypic data (217 markers) was processed by the STRUCTURE 2.3.3 software (Pritchard *et al.* 2000), applying the admixture model, a burn-in of 10,000 iterations and a 10,000 MCMC (Markov Chain Monte Carlo approach) duration to test for a Q value in the range from 1 to 15 using 15 replicates each. The likely number of sub-populations (Q) present was estimated by a phylogenetic tree based on the ICARDA DArT marker and unweighted pair group method with arithmetic average (UPGMA) using PHYLIPS 3.69 (Felsenstein 2009).

The software program TASSEL 2.1 (Bradbury *et al.* 2007) was used to calculate associations between the markers and each trait, employing the general linear model (GLM) based on the principal coordinate analysis (PCoA) by Price *et al.* (2006). The mixed linear model (MLM) suggested by Yu *et al.* (2006) was additionally implemented using PCoA and the kinship-matrix implemented by TASSEL. The PCoA was realised by calculating Eigenvalues of 702 markers by Manhattan distance and covariance. First principle coordinates (PCos) explaining in total more than 50% variance were applied. The final marker-trait association was accepted if both models (GLM and MLM) were statistically significant (P<0.05) and at least two of six possible MTAs (I1, AA1, CD1, I2, AA2, CD2) were associated with the same marker.

Annotation of resulting DArT-Markers was performed by BLASTX (Basic Local Alignment Search Tool) similarity search against the public non-redundant protein database from the National Center for Biotechnology Information (http://blast.ncbi.nlm.nih.gov/) using sequence information available on the Triticarte website (www.diversityarrays.com/sequences.html). Candidate orthologs were defined as those hits with best high scoring pair (score) and a significant E-value (Expect value) of < 1.0E-4. A consensus map created by Wenzl *et al.* (2006) revealed closely located non-DArT markers within 5 cM on either side of MTA loci which were used to identify putative functions of the associated markers available in the GrainGenes (http://wheat.pw.usda.gov) database and the Wenzl *et al.* (2006) collection of agricultural traits.

Genotypic variability

Seeds from 50 genebank accessions of *Hordeum vulgare* L. convar. *distichon* (L.) Alef. var. *nutans* (Rode) Alef. (25 accessions) and *Hordeum vulgare* L. convar. *vulgare* var. *hybernum* Viborg (25 accessions), originating from 15 countries on 4 continents were obtained from IPK Gatersleben, Germany. Each accession had been grown and harvested at IPK Gatersleben in 1974 and stored at 0±1°C in glass jars containing silica gel. Results for vigour of all accessions were available from two test years (1976 and 1993/94) and all accessions were re-tested in 2009. Changes in vigour and relationships between traits were compared by Friedman test for dependent samples, spearman correlation (r), probit analysis and ANOVA using PASW Statistic 18 (2009, Illinois, USA).

Biochemical characterisation of metabolites

Seeds from 26 genebank accessions of *Hordeum vulgare* L. convar. *distichon* (L.) Alef. var. *nutans* (Rode) Alef. (19 accessions) and *Hordeum vulgare* L. convar. *vulgare* var. *hybernum* Viborg (7 accessions) from 18 countries were multiplied at the IPK Gatersleben estate between 1995 and 2003 and in 2008. Freshly harvested seeds with a MC of approximately 6% were stored either in cold storage (CS; at 0±1°C) or at ambient storage (AS; at 20.3±2.3°C) until 2009.

Six accessions that had been multiplied in 2008 (M08) were selected for a CD experiment at two different seed MCs. Five replicates of 100 dry seeds with an initial vigour of 79.4±2.3% were equilibrated to 13% or 18% MC (Hampton and TeKrony (1995) and sealed in aluminium bags. Seeds at 18% MC were aged at 44±0.5°C for 1, 2, 3, 5 and 7 days and seeds with 13% MC for 5, 9, 13, 31 and 41 days to achieve seed vigour of approximately 60, 40, 20 and 0%. A final time interval (7 and 41 days, respectively) was chosen to produce a second measurement of 0% vigour to represent seeds aged beyond the initial vigour loss.

Fifty seeds derived from the CS, AS, M08 and CD were used for germination tests as described above. The remaining seeds were sealed in aluminium bags, frozen in liquid nitrogen and freeze-dried for 7 days. Seed MC was determined before and after freeze drying. Seeds were ground in a frozen capsule by micro-dismembrator (Retsch MM400, Germany), and the powder used for the analysis of GSH and its redox state, tocochromanols, nitrogen, starch, oil and protein content.

GSH, GSSG and other low molecular weight thiols (with brackets) and disulphides: cyst(e)ine, cyst(e)inyl-glycine and γ-glutamylcyst(e)ine were extracted and analysed by reversed-phase HPLC (Jasco, Great Dunmow, UK) as described by Kranner and Grill (1996). $E_{GSSG/2GSH}$ was calculated using the Nernst equation (Schafer and Buettner 2001, Kranner *et al.* 2006). Seed MC was used to estimate molar concentrations of GSH and GSSG, and intracellular pH was initially assumed to be 7.3 (Kranner *et al.* 2006).

Tocochromanols were isolated and separated following an adapted procedure (Bagci *et al.* 2004). 15 mg of ground seeds were extracted in 1 ml of ice-cold heptane (Fisher Scientific, Leicestershire, UK) and centrifuged at 13,000 g and 4°C for 20 min. The pellet was resuspended in 1 ml heptane and centrifuged as above. The supernatants were combined and centrifuged immediately prior to HPLC analysis. α-, β-, γ- and δ- Tocopherols and tocotrienols were separated by isocratic normal-phase HPLC (Jasco, Great Dunmow, UK), using a Supelcosil LC-Diol column [Supelco Analytical, Sigma-Aldrich from Bellefonte, USA, 250 x 4.6 mm internal diameter] and heptane : tert-butyl methyl ether (Acros Organics, New Jersey, USA) 97.5% : 2.5% (v/v) as a mobile phase at a flow rate of 1 ml min^{-1}. Tocochromanols were detected with a fluorescence detector (excitation: 295 nm; emission: 325 nm) and identified and quantified using standards (Sigma Aldrich, Poole, Dorset, UK) prepared in heptane.

Total nitrogen (TN) and total carbon (TC) content were measured in dry powdered seeds, using an electronic microbalance (M2P, Sartorius, Göttingen, Germany), by elemental analysis (Vario EL3, Elementaranalysesysteme, Hanau, Germany). Total protein was calculated as TN* multiplied by the conversion factor 5.7 for wheat grains (Sosulski & Imafidon 1990). Total lipid was analysed in powdered dry samples using NMR (MQ-60, Bruker, Rheinstetten, Germany) according to the manufacturer's instructions. For determination of starch, 10 mg of powdered samples were extracted three times with 1 ml 80% ethanol at 80°C. The remaining insoluble material was dried, and 1.5 ml 2M HCl was added to hydrolyse starch. After boiling for 1 h, the mixture was centrifuged (14,000 g, 10 min), and 5 µl of the supernatant was used for the spectrophotometric determination of glucose according to Bergmeyer *et al.* (1974).

Results

Association mapping

The genetic relationship was examined by UPGMA cluster analysis which revealed five major clusters (A to E) and several sub-clusters (Tab. S1) containing genotypes from the same germplasm pools as described by Varshney *et al.* (2010).

Processing 1 to 15 Q groups by STRUCTURE software, the population structure of the ICARDA subset consisted of four subgroups including 156 of a total 175 accessions (Fig. 1). The groups Q4 and Q2 matched with phylogenetic tree clusters D and E (Tab. S1) completely while Q1 and Q3 partially covered clusters A, B and C. The principal subgroup Q1 consisted of 22 lines belonging to two-rowed landraces from North East Asia (NEA) and Middle East Asia (MEA). Six-rowed landraces and cultivars (42 lines) from Africa (AFR) were assigned to the second group (Q2). The 55 accessions of Q3 group were mainly six-rowed landraces from NEA and the Arabian Peninsula (APS). The last subgroup (Q4) was a mixture of mainly two-rowed cultivars from all geographic regions.

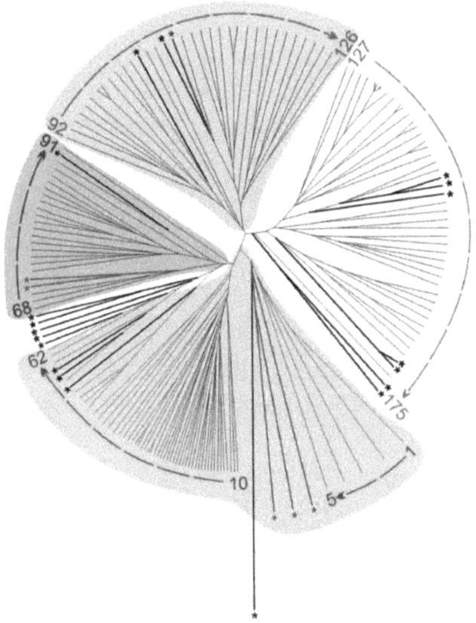

Fig. 1: Phylogenetic tree and population structure of 175 genotypes (ICARDA population) of two- and six-row barley (122 landraces and 53 cultivars) from four continents. Consecutive numbered genotypes (1-176) are described at Tab. S1. Q groups corresponding to the colour code as per figure. Black asterisks and bold lines are not clustered to any group, red and blue asterisks and bold lines are clustered to the corresponding red and blue Q group.

The PCoA, based on decomposition of any multidimensional distance metric and similar to the principal component analysis (PCA), was applied on 166 ICARDA accessions which had a MAF > 20%. The first two PCos with an individual proportion of 13.5 and 6.7% (Fig. 2) showed no definite clusters. Separate clusters were extractable once the population structure was integrated and visualised groups of Q1 to Q4. Ungrouped accessions of Q-matrix were widely distributed. In total, 13 PCos gave a proportion of >50% and were processed in association mapping analysis together with initial vigour/TG and ageing treatments (AA, CD) of the two experimental fields.

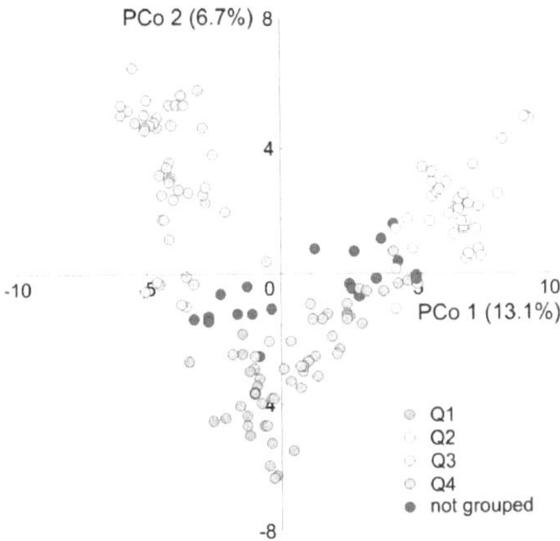

Fig. 2: **First two Principle Coordinates (PCos) of 166 ICARDA genotypes.** Associated groups of population structure are highlighted by colour code as per figure.

A maximum of 152 lines of field 1 population and 160 lines of field 2 population (seed supply of remnant lines were not sufficient) resulted in an average initial vigour of 71.8±13.8% (field 1) and 87.0±8.2% (field 2; Fig. 3). AA (at 100% RH) resulted in the depletion of vigour to 10.4±10.1% and 43.5±19.1% for field 1 and field 2, respectively. CD (at 60% RH) was less destructive, resulting in higher vigour of 51.7±19.5% and 69.4±16.4%. TG was significantly different to vigour (r=0.964, P<0.001) and both were significantly different between field 1 and field 2. Relationships between country of origin, variety, rows, Q-groups and vigour were not detected.

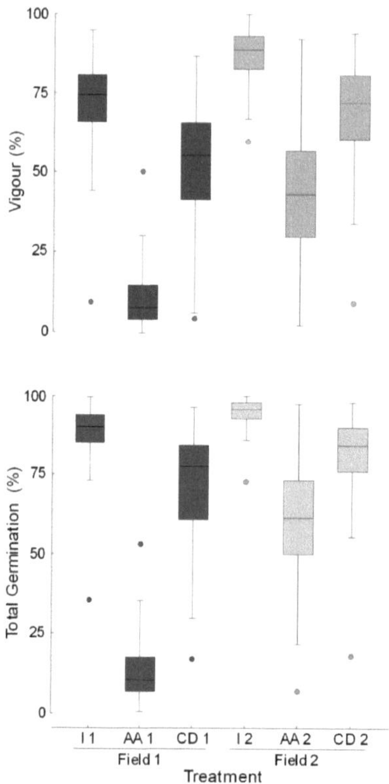

Fig. 3: Effects of different ageing treatments on seed germination across 175 genotypes. Seeds of the 175 ICARDA genotypes were multiplied in 2008 and plants were grown in two separate experimental fields in the Gatersleben estate (dark green or dark blue represents field 1, light green or light blue represents field 2). Green, vigour; blue, total germination; I, initial vigour/ total germination of the freshly harvested seeds in 2009/10; AA, accelerated ageing at 45 °C and 100% RH for 3 d; CD, controlled deterioration at 45 °C and 60 % RH for 15 d. Boxplots represent median (black line), lower and upper quartile (box), whiskers are the 98th percentile and extreme outliers (dot).

Analysis of data from control, AA and CD treatments of population lines multiplied on field 1 and 2 gave a total of 105 MTAs on 32 loci for vigour and a total of 98 MTAs on 28 loci for TG, of which six and three, respectively, were highly significant (R^2>0.06; P<0.001; Fig. 4). Highest numbers of MTAs were presented by chromosome 7H (10 MTAs), followed by 6H (8 MTAs) and 2H (8 MTAs) in the case of vigour and 2H (10 MTAs) followed by 7H (8 MTAs) and 6H (7MTAs) in the case of TG. Twenty-one loci were identical between vigour and TG. Considering only vigour, multiplication specific loci bPb4809_5H for field 1 and bPb8143_2H, bPb6868_7H for field 2 or ageing specific loci bPb4389_7H for AA and bPb7588_2H, bPb4699_4H and bPb0029_5H for CD assumed a close relationship to environmental conditions. Initial germination quality strongly influenced seed longevity in 18 of 32 vigour loci, particularly in material of field 1 which appeared in 13 of 18 loci. Most MTAs appeared in CD treatment of seed from field 1 (vigour=30; TG=24) and AA of seed from field 2 (vigour=19; TG=17), which had vigour and TG results with the highest SD and showed dependency of a wide range of genotypes.

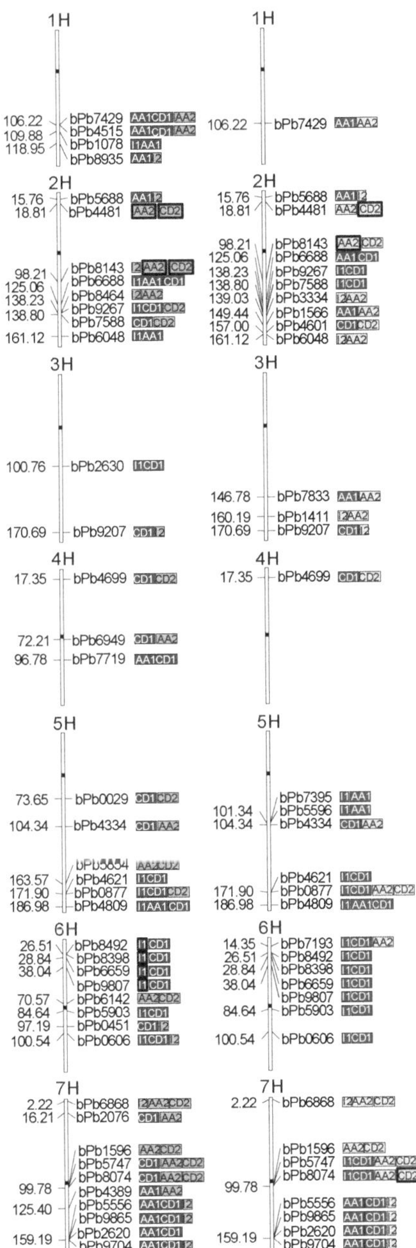

Fig. 4: Marker-trait association (MTA) across 175 genotypes (ICARDA population) in relation to seed ageing. Total germination (TG, blue) and vigour (green) expressed as TG minus abnormal seedlings are shown for the seven barley chromosomes. Dark blue and dark green indicate seeds harvested from experimental field 1, and light blue and light green show those from field 2. MTAs were identified at P<0.05, bold framed MTAs were highly significant (P<0.001) regarding two analyses methods (MLM, GLM). Abbreviations as in Fig. 2. Loci on the chromosomes are shown when at least two MTAs were detected for one locus across the six treatments. Centromers are indicated by black squares within chromosomes.

Genotypic variability

Fifty genebank accessions, belonging to two- and six-rowed barley of *Hordeum vulgare* L. from Asia, Europe and Northern Africa, showed a significant decline (P<0.001) in vigour after 35 years of storage (Fig. 5). In 1976, initial vigour was 98.5±2.2% and decreased after 19 and 35 years to 93.6±3.6% and 80.1±11.1%, respectively. In 2009, HOR 1320 had the highest vigour (95.4%) and HOR 4252 the lowest (43.1%). Using probit analysis, the half-viability period (P50) was estimated to be 52.6±0.01 years. Seed RH ranged from 19.0% to 72.9% and was not related to TG. A significant correlation was shown between vigour results of 1976 and 1994 (r =0.29; P<0.05) and a high significant difference (P<0.001) appeared between barley varieties (*distichon, hybernum*) and water activity.

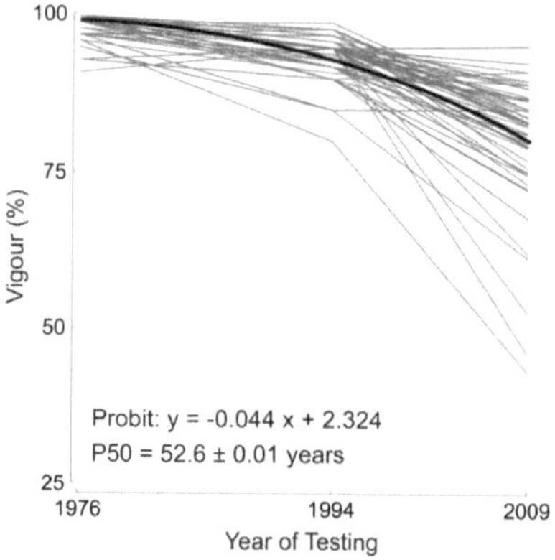

Fig. 5: Genotypic differences in seed longevity. Fifty genotypes (grey lines) were harvested in 1974 and stored at 0°C (Gatersleben barley collection). The first germination test was performed in 1976, then in 1994 and 2009. The bold black line shows a fitted probit curve across all genotypes with a P50 value of 52.6 years.

Biochemical characterisation of metabolites in relation to genotypic variation, environmental and seed storage conditions

Twenty-six barley (two- and six-rowed) accessions deriving from Africa, Europe, Asia and Central America were stored under CS and AS, and of these, six genotypes were subjected to CD at 13 and 18% MC. Genotypes 1, 5 and 6 achieved highest vigour/TG under CS, but after 14 years of AS, genotypes 2 and 4 showed highest vigour/TG. The vigour/TG of genotype 3 was strongly reduced after CS, AS and CD. However, due to the different harvest years, no clear separation between long- and short-lived genotypes was visible during CS and AS. During CD, vigour temporarily increased. Genotypes 2 and 3 tended to have shorter longevity than genotypes 1 and 5 despite similar initial vigour and TG (Fig. 6).

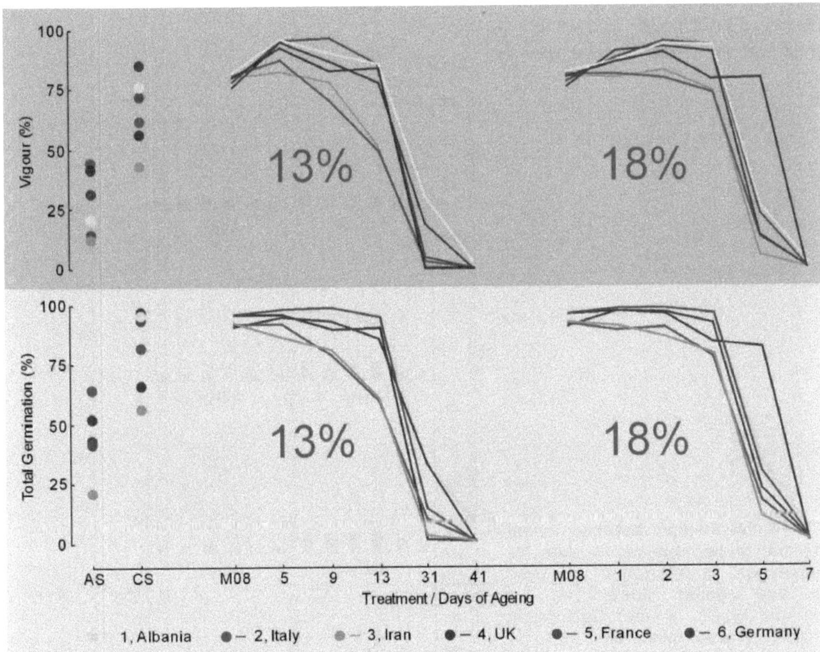

Fig. 6: Germinability of seeds in response to different storage and ageing treatments. Seeds from six genotypes of the Gatersleben barley collection from Albania (1, yellow), Italy (2, red), Iran (3, orange), UK (4, blue), France (5, green) and Germany (6, violet) were tested after ambient storage (AS) at 20°C and cold storage (CS) at 0°C (Harvest years: 1=1996, 2=1999, 3=1998, 4=2000, 5=1998, 6=1996; seeds were selected according to their high vigour). In 2008, seeds were multiplied from CS material stored at 0°C and subjected to controlled deterioration (CD13%) at 45°C, 13% seed MC for 5, 9, 13, 31 and 42 d and CD18% at 18 % seed MC for 1, 2, 3, 5 or 7 d. M08, non-aged control; Vigour (green) shows only normal seedlings; Total germination (blue) includes normal and abnormal seedlings.

A general depletion of total GSH with decreasing vigour/TG was observed (Fig. 7, 8). Significant correlations were found between GSH and vigour/TG of CD aged seeds (Tab. 1) and between GSSG and seeds from AS and CS. In CD aged seeds, the total level of GSSG was higher at 13% MC than at 18% MC but differed in addition to total glutathione between genotypes. In all treatments, genotype 6 showed significantly higher levels of total glutathione than genotype 1 despite both being long-lived accessions. Genotype 5 maintained a moderate amount of total glutathione with high TG.

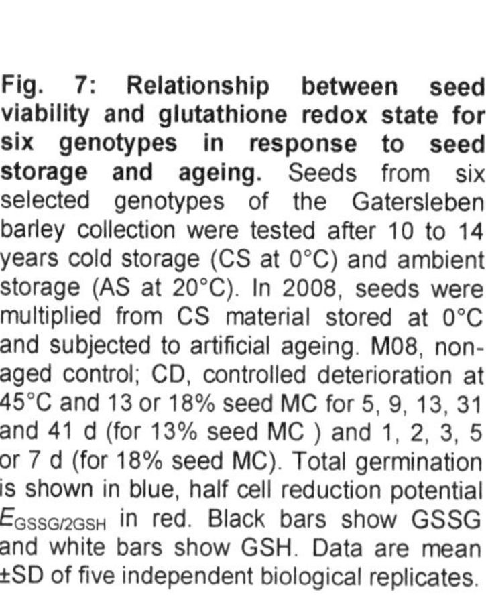

Fig. 7: Relationship between seed viability and glutathione redox state for six genotypes in response to seed storage and ageing. Seeds from six selected genotypes of the Gatersleben barley collection were tested after 10 to 14 years cold storage (CS at 0°C) and ambient storage (AS at 20°C). In 2008, seeds were multiplied from CS material stored at 0°C and subjected to artificial ageing. M08, non-aged control; CD, controlled deterioration at 45°C and 13 or 18% seed MC for 5, 9, 13, 31 and 41 d (for 13% seed MC) and 1, 2, 3, 5 or 7 d (for 18% seed MC). Total germination is shown in blue, half cell reduction potential $E_{GSSG/2GSH}$ in red. Black bars show GSSG and white bars show GSH. Data are mean ±SD of five independent biological replicates.

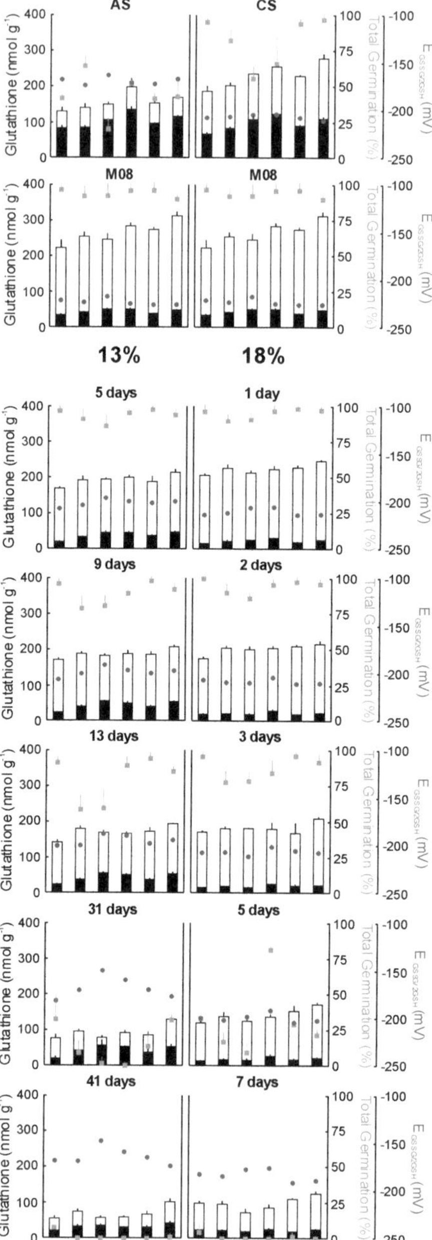

Cyst(e)ine did not correlate with vigour or TG (Tab. 1), and except for a significant (P<0.001) depletion at AS, the total cyst(e)ine was constant in all treatments (Fig. 8). The highest amounts were detected in genotype 2 and 4 (Fig. S1).

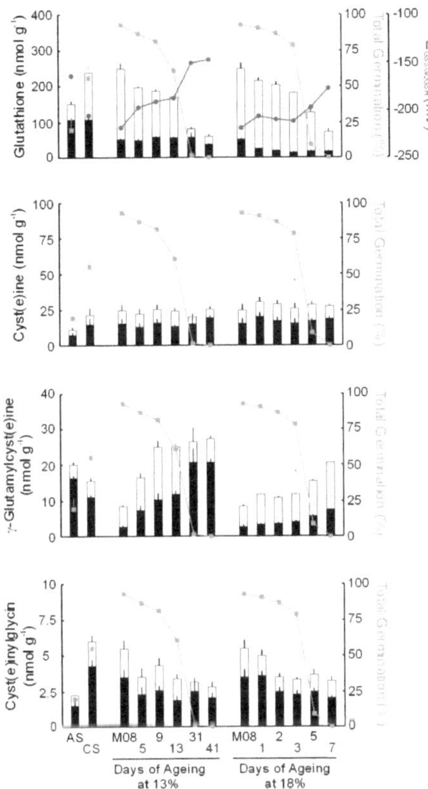

Fig. 8: **Relationship between total germination and glutathione, cyst(e)ine, γ–glutamylcyst(e)ine, cyst(e)nylglycine redox state for genotype 3 in response to seed storage and ageing.** Seeds of genotype 3 were tested after 10 to 14 years cold storage (CS at 0°C) and ambient storage (AS at 20°C). In 2008, seeds were multiplied from CS material stored at 0°C and subjected to artificial ageing. M08, non-aged control; CD, controlled deterioration at 45°C and 13 or 18% seed MC for 5, 9, 13, 31 and 41 d (for 13% seed MC) and 1, 2, 3, 5 or 7 d (for 18% seed MC). Total germination is shown in blue, half cell reduction potential $E_{GSSG/2GSH}$ in red. Black bars show disulphides (oxidised) and white bars show thiols (reduced). Data are mean ±SD of five independent biological replicates.

Highly significant correlations were found between vigour/TG and γ-glutamylcysteine in seed subject to AS and CD at 18% and 13% MC (Tab. 1). γ-Glutamylcystine increased at a higher rate in 13% MC seed and the total level tended to decrease when vigour/TG decreased to 0% (Fig. 8, S2).

In contrast, cyst(e)inylglycine decreased in content during CD and AS (Fig. 8, S3) and was positively correlated to vigour/TG at 13% MC CD and AS (Tab. 1).

Tab. 1: Correlation coefficient after spearman between vigour/Total germination (TG) of ageing treatments and thiols/disulphides, percentage GSSG (GSSG%) and the half-cell reduction potential ($E_{GSSG/2GSH}$). Seeds from 26 selected genotypes of the Gatersleben barley collection were tested, using 5 independent biological repetitions after 10 to 14 years cold storage (CS at 0°C) and, ambient storage (AS at 20°C) for the same period. In 2008, seeds were multiplied from CS material stored at 0°C and 6 of 26 genotypes were subjected to artificial ageing. M08, non-aged control; CD, controlled deterioration at 45°C and 13 or 18% seed MC for 5, 9, 13, 31 and 41 d (for 13% seed MC) and 1, 2, 3, 5 or 7 d (for 18% seed MC). Bold numbers were identified at P<0.05, red numbers were highly significant (P<0.001).

	CD 18%		CD13%		AS		CS		M08
	Vigour	TG	Vigour	TG	Vigour	TG	Vigour	TG	Vigour
Glutathione	0,74	0,83	0,78	0,78					-0,93
GSSG					-0,47	**-0,56**	-0,46	**-0,55**	
Cysteine	-0,45	-0,37	-0,36	-0,37					
Cystine									
Cysteinylglycine			**0,51**	**0,49**	0,41	0,39			
Cystinylglycine		0,42	0,67	0,65					
γ-Glutamycysteine	**-0,80**	-0,84	-0,86	-0,85	-0,62	-0,61			
γ-Glutamycystine	-0,73	-0,77							
GSSG%	-0,68	-0,68	-0,85	-0,85	**-0,54**	**-0,50**	**-0,59**	-0,63	
$E_{GSSG/2GSH}$	-0,77	-0,81	-0,90	-0,89	**-0,56**	-0,48	**-0,54**	**-0,58**	

In all seeds, $E_{GSSG/2GSH}$ was between -150 and -250mV. By plotting $E_{GSSG/2GSH}$ against %GSSG, it showed that different treatments formed clusters. $E_{GSSG/2GSH}$ and %GSSG were highly correlated with vigour/TG (Tab. 1). Oxidative stress is assumed to cause acidification of the cytoplasm. When the decrease in pH is taken into account in the Nernst equation, $E_{GSSG/2GSH}$ becomes more positive (Fig. 9 A, Inset).

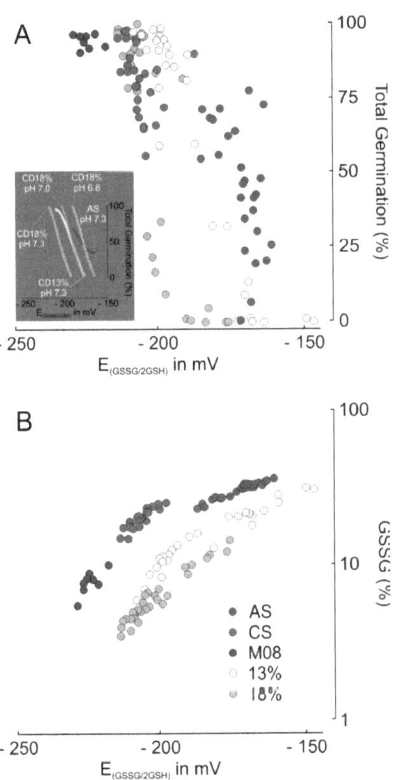

Fig. 9: Relationship between total germination and glutathione half-cell reduction potential ($E_{GSSG/2GSH}$) for 26 genotypes after storage and artificial ageing treatments. (A) $E_{GSSG/2GSH}$ was calculated from data shown in Fig. 6 (artificial ageing at 13% MC in yellow and 18% in blue, M08, non-aged control in green) and are plotted against total germination together with data from an additional 20 genotypes after cold and ambient storage (CS (red) and AS (violet)). Inset: pH value of CD18% treatment used in the Nernst equation (pH 7.0) for the calculation of $E_{GSSG/2GSH}$ was decreased and overlaps with CD13% at pH 7.0 and AS at pH 6.8 (B) $E_{GSSG/2GSH}$ plotted against GSSG as a percentage of total glutathione (GSH + GSSG) shows that the different treatments cluster.

54

Tocochromanols were predominately present as tocotrionols (Fig. 10 B) but also tocopherols (Fig. 10 A) did not show any relationship with seed viability. Similarly, oil and starch content did not significantly change with storage or ageing treatments (Fig. 10 C). The total protein content differed significantly only between ageing treatments and AS, as the CS material was derived from a different harvest year.

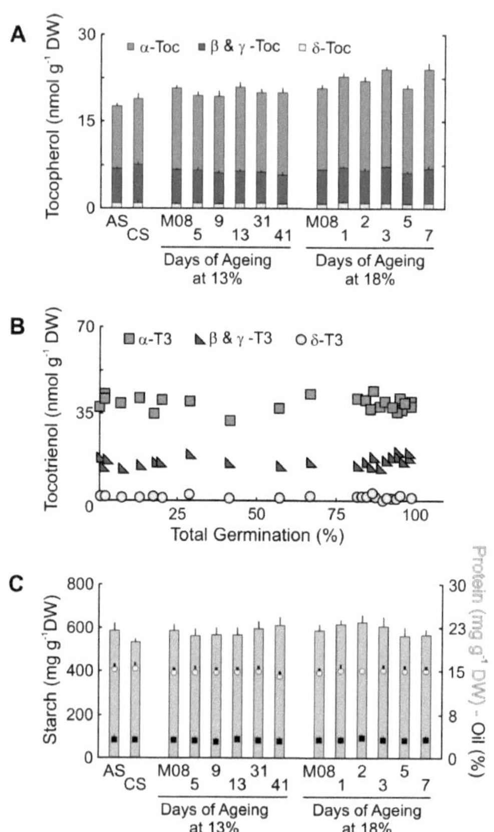

Fig. 10: Relationship between total germination, tocochromanols and seed storage reserves contents. (A) α, β, γ and δ- Tocopherol (Toc) contents are shown for genotype 3. (B) α, β, γ and δ- Tocotrienols (T3) are adjusted to tocopherol standards and shown for different total germinations. (C) Seed storage reserves of protein (yellow dot), oil (black dot) and starch (grey bars) did not change significantly in response to any treatment.

Discussion

The basic processes and influencing factors of seed longevity have been investigated since Haberlandt's experiments, but several principles remain unclear. Our study is intended to give a better understanding of the influence of genetics, environment and storage conditions on barley seed ageing.

ICARDA clusters and population structure

For genetic characterisation, an association mapping study was performed using a subset of the ICARDA germplasm collection which has been already described by Varshney *et al.* (2010). Based on a genetic similarity coefficient matrix using single nucleotide polymorphism (SNP) and simple sequence repeat (SSR) markers, genotypes from the same geographic regions were separated into clusters. *H. vulgare* accessions of the current study were identical to genotypes clustered from NEA, AFR, MEA and APS. However, there were also clusters which did not match with a specific geographic region and indicate either an artificial transfer from one region to another or an independent development of similar genetic variation (Huang *et al.* 2002, Varshney *et al.* 2010). Furthermore, Varshney's SNP and SSR markers revealed a higher number of alleles per locus in *H. spontaneum* genotypes. Due to a limited seed supply we disregarded this material but did not lose genetic information as landraces in particular showed unique alleles (Varshney *et al.* 2010).

H. vulgare genotypes were classified by DArT markers which offer a high scoring reproducibility of at least 99.9% and a genotype call rate of 98% (Wenzl *et al.* 2004, Zhang *et al.* 2009), and create a denser genetic map comprising 703 markers with average distances of 1.6 cM. The ability of DArT markers to group barley accessions according to their known relationship (Wenzl *et al.* 2004) were confirmed by the similarity between this and the Varshney *et al.* (2010) dendrogram. With some exceptions, genotypes were grouped into landraces from NEA & MEA (Q1), AFR (Q2), NEA & APS (Q3) and cultivars of different countries (Q4). Additionally, two-rowed and six-rowed barley clustered mainly in Q1 and Q4, and Q2 and Q3, respectively. This implies that mutations of the *vrs* inflorescence genes which resulted in the formation of six-rowed barleys appeared at different times and locations independently (Komatsuda *et al.* 2007) but may have focused on the areas Africa, NEA and the APS.

ICARDA multiplications and genotype effects

Seeds of two ICARDA multiplications were exposed to two artificial ageing methods and the germination results were employed for association mapping. Vigour/TG results of different ICARDA genotypes as same as accessions after 25 years of storage separated over a wide range vigour/TG percentages which indicates a genetic component of seed longevity. Further seed from field 2 showed significantly higher vigour and TG both initially and after AA and CD in comparison with field 1 (Fig. 3), illustrating that environmental and developmental conditions also affect the germination (Koornneef *et al.* 2002). The differences in seed between the two fields could be due to a variety of soil properties (Tab. 2).

Tab. 2: Soil properties of experimental field used for ICARDA multiplication. Investigation of soil characteristics was performed by AGROLAB Boden- und Pflanzenberatungsdienst GmbH, Burgstraße 57, 99986 Oberdorla, Germany at 27th March 2008.

	Field 1	Field 2
Soil type	Clay silt (toniger Lehm)	Humous clay silt (humoser toniger Lehm)
pH	7.6	7.3
Phosphor ($mg \cdot 100g^{-1}$)	6.9	17.6
Potassium ($mg \cdot 100g^{-1}$)	16.6	23.2
Magnesium ($mg \cdot 100g^{-1}$)	9.8	17.1
Organic matter (%)	-	4.0
NO_3 ($kg\ N \cdot ha^{-1}$)	14	28
NH_4 ($kg\ N \cdot ha^{-1}$)	2	2
Mineral N ($kg\ N \cdot ha^{-1}$)	16	20

The initial vigour has a strong influence on seed longevity (Roberts 1973). Therefore, initial vigour/TG were also integrated into the association analyses and resulted in 26 of a total 105 MTAs related to seed ageing (Fig. 4). Initial vigour, TG and ageing treatment were not dependent on geographical origin, Q group, rows or history (cultivar/landraces), also supported by the 50 genotypes in Fig. 5, meaning these traits can be excluded as influencing factors on seed longevity.

Marker-Trait Associations and its possible functions

Markers resulting from the association mapping were compared to the consensus map of Wenzl *et al.* (2006) and revealed 11±6 DArT markers and 4±3 non-DArT markers within 5 cM on either side of MTA loci. Previously reported non-DArT markers were linked with agronomic traits and identified putative functions. BLASTX search supported further assumptions.

Present identified barley longevity loci have been found in vicinity with QTLs of a previous study of Nagel *et al.* (2009). At the distal region of chromosome 2H the expressed sequence tag (EST) markers of the Oregon Wolfe Barley (OWB) population were associated with dehydration-responsive element binding (DREB) proteins which play a crucial role in the plant response to drought, salt and cold stress in rice (Dubouzet *et al.* 2003). Barley genes involved in the response to drought and low temperature were additionally found by Rostoks *et al.* (2005).

Barley QTLs related to pathogen response to *Fusarium graminearum* (Zhu *et al.* 1999) and the cereal cyst nematode (Kretschmer *et al.* 1997) are found in a more central region of 2H close to the florescence gene *Vsr1*. This gene might be involved in longevity traits but did not show any relationship with the row-characteristics of the spikes. Other QTLs involved in pathogen response to *Pyrenophora graminea* which is close to bPb_4699 (Pecchioni *et al.* 1996), and stem rust, scald and net blotch which are close to bPb_7719 (Spaner *et al.* 1998), have a massive impact on the malt quality and are located on 4H. Salt tolerance mechanisms also appear to be closely connected to bPb_4699 (Mano & Takeda 1997). On the same chromosome, a region for aluminium tolerance was detected by Navakode *et al.* (2009) which showed sequence homology to a linked vascular proton ATPase A3.

Quantitative trait loci related to germination e.g. abscisic acid (ABA), germination rate (bPb_0029; Mano & Takeda 1997), dormancy (bPb_4621; Romagosa *et al.* 1999) and preharvest sprouting (bPb_4809; Li *et al.* 2003) are located on 5H. The hormonal balance between gibberellic acid (GA) and ABA regulate dormancy release and germination (North *et al.* 2010). ABA signals are transduced either by transcription factors such as ABI3 (Holdsworth *et al.* 2008) which is involved in the expression of maturation genes, seed storage protein genes (Kroj *et al.* 2003), *LEA* genes (Giraudat *et al.* 1992), genes with antioxidant functions (Haslekas *et al.* 2003) and heat shock protein genes (Kotak *et al.* 2007) or by protein kinases. A serine/threonine-specific protein kinase was identified in a BLASTX search using bPb_4334 and is described as essential for biotic stress resistance against stem rust (Brueggeman et al. 2008). At the same position previous artificial ageing experiments with the Steptoe-Morex population of barley revealed a QTL for seed longevity with ESTs that were related to heat shock proteins. These also appear on the second seed longevity QTL of the OWB population in the distal region of 5H (Nagel *et al.* 2009). Additionally, the transcription factor APETALA2 which usually determines floral meristem identity (Chuck *et al.* 1998) seems to be connected with seed longevity. Furthermore, APETALA2 influences the development of the major seed compartments of the embryo, endosperm and seed coat (Ohto *et al.* 2009) which performs an important function to protect the embryo and seed reserves from abiotic and biotic stress (Debeaujon *et al.* 2000, Rajjou & Debeaujon 2008, North *et al.* 2010).

Biotic stress resistance genes were found after a BLASTX search for bPb_8492, bPb_8398, bPb_6659 and bPb_9807 and included a serine/threonine-specific protein kinase and NBS_LRR disease resistance proteins, characterized by an N-terminal nucleotide-binding site (NBS) and C-terminal leucine rich repeats (LRRs). Both work together to provide resistance to plant pathogenic organisms (Brueggeman *et al.* 2008). One of the resistance genes against the disease net blotch (Steffenson *et al.* 1996) co-localised at this position. Further loci related to net blotch are bPb_5903 (Cakir *et al.* 2003) and bPb2076 (Steffenson *et al.* 1996) which is also co-localised with a QTL for *Fusarium graminearum* (Zhu *et al.* 1999) and stem rust (Brueggeman *et al.* 2002) on 7H. On chromosome 7H, the third seed longevity QTL of OWB population appears in the region of bPb5747 and is linked with ethylene-responsive element (ERE) binding proteins which have a significant role in plant adaptation to abiotic stress (Nagel *et al.* 2009), and may be regulated by ABA-dependent and −independent pathways under conditions of cold, dehydration and high salinity (Yamaguchi-Shinozaki & Shinozaki 2006). The ERE binding protein family belongs to the ERF/APETALA2 superfamily and are involved in the plant response to multiple stresses including up-regulation of antioxidant activity (Tang *et al.* 2005) which leads to decreased ROS formation and enhanced resistance to drought, salt and freezing stress (Wu *et al.* 2008).

The distal region of 7H bPb_5556 is related to sucrose synthase 5 and 6 which are widely believed to be the main route of entry of carbon from sucrose into cellular metabolism in plants. These isoforms are closely related in *Arabidopsis* but have different spatial and temporal patterns of expression. Some react strongly to hypoxia indicating a connection with critical ATP formation under oxygen limitation (Bieniawska *et al.* 2007).

Interestingly a higher number of vigour MTAs are associated with pre-haverst conditions, indicating that environmental conditions upon maturation can harm seeds and influence the initial seed quality, although they may not prevent germination. Additionally the majority of vicinity QTLs linked with germination rate (Mano & Takeda 1997) were associated with vigour indicating a close relationship between the appearance of normal seedlings and their speed of emergence.

The connection of longevity loci to abiotic and biotic stress resistance indicated pleiotropic effects and confirmed the assumption of an overall tolerance mechanism to various seed stresses (Clerkx *et al.* 2004). In some cases MTAs were not co-located with other traits which might show ageing specific loci. Further QTL detection was concentrated on agronomic traits such as pathogen-resistance, yield and plant growth parameters. The identification of these QTLs in relation to longevity symbolises a close connection between environmental conditions during development, chemical composition and the ability to survive storage. Unfortunately we cannot distinguish between QTLs linked to environmental effects and QTLs influenced by storage.

Repair during early ageing

During initial AA at 13% and 18% MC vigour increased, indicative of the activation of repair mechanisms (Fig. 6). This was supported by the presence of a marker on chromosome 7H, connected to ERE binding proteins, linked to a DNA helicase which is involved in DNA replication, repair, recombination and transcription. Furthermore, a BLASTX search for the closely connected DArT marker bPb_6688, revealed homology to RNAseH which is ubiquitous in all prokaryotic cells and an important enzyme in biochemical processes associated with DNA replication, gene expression and DNA repair (Tadokoro & Kanaya 2009).

Reactive oxygen species and the antioxidant glutathione

The association mapping study revealed a broad spectrum of MTAs related to abiotic and biotic stresses which disrupt cellular homeostasis and enhance ROS production (Mittler 2002). ROS activate several stress response pathways (op den Camp *et al.* 2003) and low molecular weight antioxidants, such as GSH and tocopherol provide essential information on the cellular redox state and influence gene expression (Foyer & Noctor 2005). The occurrence of loci and markers related to these stresses show a stress memory which is saved in the biochemical status of the seeds.

A general depletion of the total GSH pool and a shift towards more oxidising conditions occurs during ageing (Kranner & Grill 1993, De Gara *et al.* 2003, Seal *et al.* 2010). A high concentration of GSH does not mean a high viability *per se*. Comparison of six barley genotypes over different storage and ageing treatments showed that TG was maintained in genotypes with different concentrations of total GSH and values of $E_{GSSG/2GSH}$. In poplar, experiments overexpressing γ-glutamylcysteine synthase (γ-ECS), which is necessary for GSH synthesis, resulted in a significantly higher GSH content but higher stress resistance could not be proved (Noctor *et al.* 1998). The efficient level of GSH is assumed to be genotype-specific and may depend on the pre-harvest conditions. In agreement with these findings, neither GSH nor GSSG alone correlated significantly with vigour/TG in long-term and artificially aged material. However, highly significant correlations across all treatments were found for $E_{GSSG/2GSH}$ and %GSSG.

If the GSH pool is depleted during seed storage the result will be lethal. Mutant *Arabidopsis* seeds defective in GSH biosynthesis lacked GSH in the embryo and did not germinate (Cairns *et al.* 2006). The current results show viability loss occurred before GSH was depleted, which indicates the involvement of programmed cell death (PCD) which is triggered by a decrease in the GSH/GSSG ratio (Kranner *et al.* 2006).

In dry seeds, GSH is oxidised by ROS to form GSSG and is not regenerated (Colville & Kranner 2010). The initial GSSG concentration of freshly harvested barley seeds accounted for 8% of the total GSH + GSSG pool and after a maximum of 15 years, seeds from CS and AS contained 21% and 32% GSSG, respectively. The GSSG concentration was higher in AS and CS seeds than in seeds subjected to CD, suggesting that enzymes that reduce GSSG to GSH were not active at the lower MCs of AS and CS seeds, as seen previously in pea seeds (Kranner & Grill 1993). The accumulation of γ-glutamylcyst(e)ine with vigour loss was significantly higher at AS, CS and CD at 13% MC indicating a partial degradation of GSH to the disulphide. At 18% MC, this process was less destructive and enzymes might synthesise further GSH. With the exception of AS, the cyst(e)ine pool was unaffected by ageing treatments, and may be an important reserve for GSH synthesis during seedling growth (Rauser *et al.* 1991).

TG was lowest in seeds after AS suggesting a GSSG signalling function for apoptosis by activation of mitogen-activated protein kinases (MAPK) which belong to the serine/ threonine protein kinases (Filomeni *et al.* 2005, Colcombet & Hirt 2008) and were already linked in the current association mapping study. However, the application of the MAPK concept on dry seed is problematic as protein kinases are activated by phosphorylation which requires ATP (Filomeni *et al.* 2005) and dry seeds are in a resting state where ATP availability could be limited. Additionally, Mhamdi *et al.* (2010) studied *Arabidopsis* catalase single and double mutants and observed extremely high GSSG levels without any necrotic cell death. Therefore a simple relationship between GSSG content and cell death is unlikely (Queval *et al.* 2007). Furthermore, our genetic study revealed an activation of protection and repair mechanisms so self-destruction of the cell is unlikely although accessions close to vigour loss were unavailable to confirm this.

Seed ageing and pH shift

Stress causes acidification of the cytoplasm and decreases intracellular pH (Kurkdjian & Guern 1989, Kranner et al. 2006). When decreasing pH was applied to the Nernst equation, the 18% MC ageing curve could be shifted to that of the drier storage conditions. GSSG is generated in the apoplast but is transported to the cytosol for reduction to GSH via proton

60

symport (Swanson *et al.* 1998) [confirmed by in the genetic study (vascular proton ATPase A3)] that also results in proton accumulation and acidification, however it is likely that several glutathione transport mechanisms exist in parallel as described for ascorbate (Foyer *et al.* 2001).

Tocochromanols during ageing

Tocopherols and tocotrienols, commonly known as tocochromanols (Falk & Munné-Bosch 2010), are lipophilic molecules, produced by photosynthetically active plants. They scavenge acyl peroxyl radicals and protect membranes against fatty acid peroxidation (Smirnoff 2010). Especially tocopherols are often mentioned in relation to seed longevity of Arabidopsis seed (Sattler *et al.* 2004, Mène-Saffrané *et al.* 2010), in seeds of the halophyte *Suaeda maritima* (Seal *et al.* 2010) and in beech seeds (*Fagus sylvatica*) (Ratajczak & Pukacka 2005). However, the present study with barley could not confirm any relationship between seed longevity and tocochromonals. Neither the most abundant tocotrienols nor tocopherols change significantly during long-term storage or artificial ageing assuming another role than as an antioxidant in barley seed (Falk & Munné-Bosch 2010).

Storage compounds during ageing

Similar to the tocochromanols, the main storage compounds starch, protein and oil did not show any alteration during ageing. This is in agreement with the finding that no correlations between storage compounds (total sugar, oligosaccharides, sucrose content, soluble proteins) of 118 species and their survival were found (Walters *et al.* 2005b) indicating that not a general reserve depletion but the failure to mobilise reserves correlates with viability loss (Kranner *et al.* 2010b). By contrast Walters *et al.* (2005b) and also Nagel *et al.* (2010) observed slightly significant correlation of lipid content and viability, raising further questions regarding putative relationships between seed longevity and lipid stability. However, in several legume species this aspect could not be proved (Ponquett *et al.* 1992).

Conclusions

This chapter demonstrates that seed deterioration is influenced by genetic, environmental and biochemical factors. Loci detected for artificial ageing were related to agronomic traits and abiotic and biotic stress response. The identification of these QTLs in relation with longevity studies show that a close connection exists between environmental conditions during development, the chemical composition and the ability to survive storage. Hence, germinability and seed ageing are rather complex traits. The assumed involvement of ROS was confirmed by the degradation of a major antioxidant GSH. The response of GSH differed in artificially aged and in long-term stored seeds. This needs to be considered for future research into the biochemical processes of viability loss during long-term storage.

Acknowledgements

The authors would like to thank the staff of the federal *ex situ* genebank of Germany for providing seeds, particularly Sibylle Pistrick, Anita Winger, Stefanie Thumm and Mariann Börner for excellent technical assistance. For valuable comments on the experiments and the manuscript thank you to Hugh Pritchard. The Royal Botanic Gardens, Kew, receive grant-in-aid from DEFRA.

References

Bagci E., M. Vural, T. Dirmenci, L. Bruehl & K. Aitzetmüller. (2004) Fatty acid and tocochromanol patterns of some *Salvia* L. species. Zeitschrift für Naturforschung C-A Journal of Biosiences 59, 305-309.

Bailly C. (2004) Active oxygen species and antioxidants in seed biology. Seed Science Research 14, 93-107.

Bergmeyer H.U., E. Bernt, F. Schmid & H. Stark. (1974) Methods of enzymatic analysis. Verlag Chemie, Weinheim.

Bieniawska Z., D.H. Paul Barratt, A.P. Garlick, V. Thole, N.J. Kruger, C. Martin, R. Zrenner & A.M. Smith. (2007) Analysis of the sucrose synthase gene family in *Arabidopsis*. The Plant Journal 49, 810-828.

Bradbury P.J., Z. Zhang, D.E. Kroon, T.M. Casstevens, Y. Ramdoss & E.S. Buckler. (2007) TASSEL: software for association mapping of complex traits in diverse samples. Bioinformatics 23, 2633-2635.

Bray C.M. & J. Dasgupta. (1976) Ribonucleic-acid synthesis and loss of viability in pea seed. Planta 132, 103-108.

Brueggeman R., N. Rostoks, D. Kudrna, A. Kilian, F. Han, J. Chen, A. Druka, B. Steffenson & A. Kleinhofs. (2002) The barley stem rust-resistance gene *Rpg1* is a novel disease-resistance gene with homology to receptor kinases. Proceedings of the National Academy of Sciences of the United States of America 99, 9328-9333.

Brueggeman R., A. Druka, J. Nirmala, T. Cavileer, T. Drader, N. Rostoks, A. Mirlohi, H. Bennypaul, U. Gill, D. Kudrna, C. Whitelaw, A. Kilian, F. Han, Y. Sunl, K. Gill, B. Steffenson & A. Kleinhofs. (2008) The stem rust resistance gene *Rpg5* encodes a protein with nucleotide-binding-site, leucine-rich, and protein kinase domains. Proceedings of the National Academy of Sciences of the United States of America 105, 14970-14975.

Cairns N.G., M. Pasternak, A. Wachter, C.S. Cobbett & A.J. Meyer. (2006) Maturation of *Arabidopsis* seeds is dependent on glutathione biosynthesis within the embryo. Plant Physiology 141, 446-455.

Cakir M., D. Poulsen, N.W. Galwey, G.A. Ablett, K.J. Chalmers, G.J. Platz, R.F. Park, R.C.M. Lance, J.F. Panozzo, B.J. Read, D.B. Moody, A.R. Barr, P. Johnston, C.D. Li, W.J.R. Boyd, C.R. Grime, R. Appels, M.G.K. Jones & P. Langridge. (2003) Mapping and QTL analysis of the barley population Tallon x Kaputar. Australian Journal of Agricultural Research 54, 1155-1162.

Chuck G., R.B. Meeley & S. Hake. (1998) The control of maize spikelet meristem fate by the APETALA2-like gene indeterminate spikelet1. Genes & Development 12, 1145-1154.

Clerkx E.J.M., M.E. El-Lithy, E. Vierling, G.J. Ruys, H. Blankestijn-De Vries, S.P.C. Groot, D. Vreugdenhil & M. Koornneef. (2004) Analysis of natural allelic variation of *Arabidopsis* seed germination and seed longevity traits between the accessions Landsberg erecta and Shakdara, using a new recombinant inbred line population. Plant Physiology 135, 432-443.

Colcombet J. & H. Hirt. (2008) *Arabidopsis* MAPKs: a complex signalling network involved in multiple biological processes. Biochemical Journal 413, 217-226.

Colville L. & I. Kranner. (2010) Desiccation tolerant plants as model systems to study redox regulation of protein thiols. Plant Growth Regulation.

Davies M.J., S.L. Fu, H.J. Wang & R.T. Dean. (1999) Stable markers of oxidant damage to proteins and their application in the study of human disease. Free Radical Biology and Medicine 27, 1151-1163.

De Gara L., M.C. de Pinto, V.M.C. Moliterni & M.G. D'Egidio. (2003) Redox regulation and storage processes during maturation in kernels of *Triticum durum*. Journal of Experimental Botany 54, 249-258.

De Gara L., V. Locato, S. Dipierro & M.C. de Pinto. (2010) Redox homeostasis in plants. The challenge of living with endogenous oxygen production. Respiratory Physiology & Neurobiology 173, S13-S19.

Debeaujon I., K.M. Leon-Kloosterziel & M. Koornneef. (2000) Influence of the testa on seed dormancy, germination, and longevity in *Arabidopsis*. Plant Physiology 122, 403-413.

Delouche J.C. & C.C. Baskin. (1973) Accelerated aging techniques for predicting the relative storability of seed lots. Seed Science and Technology 1.

Dubouzet J.G., Y. Sakuma, Y. Ito, M. Kasuga, E.G. Dubouzet, S. Miura, M. Seki, K. Shinozaki & K. Yamaguchi-Shinozaki. (2003) *OsDREB* genes in rice, *Oryza sativa* L., encode transcription activators that function in drought-, high-salt- and cold-responsive gene expression. Plant Journal 33, 751-763.

El-Maarouf-Bouteau H. & C. Bailly. (2008) Oxidative signaling in seed germination and dormancy. Plant Signaling & Behavior 3, 175-182.

Falk J. & S. Munné-Bosch. (2010) Tocochromanol functions in plants: Antioxidation and beyond. Journal of Experimental Botany 61, 1549-1566.

FAO. (2011) FAOSTAT: Crops statistic 2009. Food and Agriculture Organization of the United Nations, http://faostat.fao.org/site/567/DesktopDefault.aspx?PageID=567#ancor, Rome, visited at 06th April 2011.

Felsenstein J. (2009) PHYLIP 3.69 (Phylogeny Inference Package), Washington.

Filomeni G., G. Rotilio & M.R. Ciriolo. (2005) Disulfide relays and phosphorylative cascades: partners in redox-mediated signaling pathways. Cell Death Differ 12, 1555-1563.

Foyer C.H. & G. Noctor. (2005) Redox homeostasis and antioxidant signaling: A metabolic interface between stress perception and physiological responses. Plant Cell 17, 1866-1875.

Foyer C.H., F.L. Theodoulou & S. Delrot. (2001) The functions of inter- and intracellular glutathione transport systems in plants. Trends in Plant Science 6, 486-492.

Gale M.D. & K.M. Devos. (1998) Plant Comparative Genetics after 10 Years. Science 282, 656-659.

Giraudat J., B.M. Hauge, C. Valon, J. Smalle, F. Parcy & H.M. Goodman. (1992) Isolation of the *Arabidopsis-Abi3* gene by positional cloning. Plant Cell 4, 1251-1261.

Halliwell B. (2006) Reactive species and antioxidants. Redox biology is a fundamental theme of aerobic life. Plant Physiology 141, 312-322.

Hampton J.G. & D.M. TeKrony. (1995) Handbook of vigour test methods. 3 ed. International Seed Testing Association, Zürich.

Harrington J.F. (1963) Practical instructions and advice on seed storage. Proceedings of the International Seed Testing Association pp. 989-994.

Haslekas C., M.K. Viken, P.E. Grini, V. Nygaard, S.H. Nordgard, T.J. Meza & R.B. Aalen. (2003) Seed 1-cysteine peroxiredoxin antioxidants are not involved in dormancy, but contribute to inhibition of germination during stress. Plant Physiology 133, 1148-1157.

Hay F.R., J. Adams, K. Manger & R. Probert. (2008) The use of non-saturated lithium chloride solutions for experimental control of seed water content. Seed Science and Technology 36, 737-746.

Holdsworth M.J., L. Bentsink & W.J.J. Soppe. (2008) Molecular networks regulating *Arabidopsis* seed maturation, after-ripening, dormancy and germination. New Phytologist 179, 33-54.

Huang X.Q., A. Börner, M.S. Roder & M.W. Ganal. (2002) Assessing genetic diversity of wheat (*Triticum aestivum* L.) germplasm using microsatellite markers. Theoretical and Applied Genetics 105, 699-707.

ISTA. (2008) International Rules for Seed Testing International Seed Testing Association, Bassersdorf, Switzerland.

Komatsuda T., M. Pourkheirandish, C.F. He, P. Azhaguvel, H. Kanamori, D. Perovic, N. Stein, A. Graner, T. Wicker, A. Tagiri, U. Lundqvist, T. Fujimura, M. Matsuoka, T. Matsumoto & M. Yano. (2007) Six-rowed barley originated from a mutation in a homeodomain-leucine zipper I-class homeobox gene. Proceedings of the National Academy of Sciences of the United States of America 104, 1424-1429.

Koornneef M., L. Bentsink & H. Hilhorst. (2002) Seed dormancy and germination. Current Opinion in Plant Biology 5, 33-36.

Kotak S., E. Vierling, H. Baumlein & P. von Koskull-Doring. (2007) A novel transcriptional cascade regulating expression of heat stress proteins during seed development of *Arabidopsis*. Plant Cell 19, 182-195.

Kranner H. & D. Grill. (1993) Content of low-molecular-weight thiols during the imbibition of pea seeds. Physiologia Plantarum 88, 557-562.

Kranner I. & D. Grill. (1996) Determination of glutathione and glutathione disulphide in lichens: A comparison of frequently used methods. Phytochemical Analysis 7, 24-28.

Kranner I., S. Birtic, K.M. Anderson & H.W. Pritchard. (2006) Glutathione half-cell reduction potential: A universal stress marker and modulator of programmed cell death? Free Radical Biology and Medicine 40, 2155-2165.

Kranner I., F.V. Minibayeva, R.P. Beckett & C.E. Seal. (2010a) What is stress? Concepts, definitions and applications in seed science. New Phytologist 188, 655-673.

Kranner I., G. Kastberger, M. Hartbauer & H.W. Pritchard. (2010b) Noninvasive diagnosis of seed viability using infrared thermography. Proceedings of the National Academy of Sciences of the United States of America 107, 3912-3917.

Kranner I., T. Roach, R.P. Beckett, C. Whitaker & F.V. Minibayeva. (2010c) Extracellular production of reactive oxygen species during seed germination and early seedling growth in *Pisum sativum*. Journal of Plant Physiology 167, 805-811.

Kretschmer J.M., K.J. Chalmers, S. Manning, A. Karakousis, A.R. Barr, A.K.M.R. Islam, S.J. Logue, Y.W. Choe, S.J. Barker, R.C.M. Lance & P. Langridge. (1997) RFLP mapping of the *Ha 2* cereal cyst nematode resistance gene in barley. Theoretical and Applied Genetics 94, 1060-1064.

Kroj T., G. Savino, C. Valon, J. Giraudat & F. Parcy. (2003) Regulation of storage protein gene expression in *Arabidopsis*. Development 130, 6065-6073.

Kuckuck H. (1929) Die Genetik der Gerste. Theoretical and Applied Genetics 2, 50-60.

Kurkdjian A. & J. Guern. (1989) Intracellular pH: Measurement and importance in cell activity. Annual Review of Plant Physiology and Plant Molecular Biology 40, 271-303.

Li C.D., R.C.M. Lance, H.M. Collins, A. Tarr, S. Roumeliotis, S. Harasymow, M. Cakir, G.P. Fox, C.R. Grime, S. Broughton, K.J. Young, H. Raman, A.R. Barr, D.B. Moody & B.J. Readf. (2003) Quantitative trait loci controlling kernel discoloration in barley (*Hordeum vulgare* L.). Australian Journal of Agricultural Research 54, 1251-1259.

Mano Y. & K. Takeda. (1997) Mapping quantitative trait loci for salt tolerance at germination and the seedling stage in barley (*Hordeum vulgare* L). Euphytica 94, 263-272.

McDonald M.B. (1999) Seed deterioration: Physiology, repair and assessment. Seed Science and Technology 27, 177-237.

Mène-Saffrané L., A.D. Jones & D. DellaPenna. (2010) Plastochromanol-8 and tocopherols are essential lipid-soluble antioxidants during seed desiccation and quiescence in *Arabidopsis*. Proceedings of the National Academy of Sciences 107, 17815-17820.

Mhamdi A., G. Queval, S. Chaouch, S. Vanderauwera, F. Van Breusegem & G. Noctor. (2010) Catalase function in plants: A focus on *Arabidopsis* mutants as stress-mimic models. Journal of Experimental Botany 61, 4197-4220.

Mittler R. (2002) Oxidative stress, antioxidants and stress tolerance. Trends in Plant Science 7, 405-410.

Miura K., S.Y. Lin, M. Yano & T. Nagamine. (2002) Mapping quantitative trait loci controlling seed longevity in rice (*Oryza sativa* L.). Theoretical and Applied Genetics 104, 981-986.

Nagel M. & A. Börner. (2010) The longevity of crop seeds stored under ambient conditions. Seed Science Research 20, 1-12.

Nagel M., H. Vogel, S. Landjeva, G. Buck-Sorlin, U. Lohwasser, U. Scholz & A. Börner. (2009) Seed conservation in *ex-situ* genebanks - genetic studies on longevity in barley. Euphytica 170, 1-10.

Navakode S., A. Weidner, R.K. Varshney, U. Lohwasser, U. Scholz & A. Börner. (2009) A QTL analysis of aluminium tolerance in barley, using gene-based markers. Cereal Research Communications 37, 531-540.

Noctor G., A.-C.M. Arisi, L. Jouanin, K.J. Kunert, H. Rennenberg & C.H. Foyer. (1998) Glutathione: Biosynthesis, metabolism and relationship to stress tolerance explored in transformed plants. Journal of Experimental Botany 49, 623-647.

North H., S. Baud, I. Debeaujon, C. Dubos, B. Dubreucq, P. Grappin, M. Jullien, L. Lepiniec, A. Marion-Poll, M. Miquel, L. Rajjou, J.M. Routaboul & M. Caboche. (2010) *Arabidopsis* seed secrets unravelled after a decade of genetic and omics-driven research. Plant Journal 61, 971-981.

Ohto M.-a., S. Floyd, R. Fischer, R. Goldberg & J. Harada. (2009) Effects of APETALA2 on embryo, endosperm, and seed coat development determine seed size in *Arabidopsis*. Sexual Plant Reproduction 22, 277-289.

op den Camp R.G.L., D. Przybyla, C. Ochsenbein, C. Laloi, C. Kim, A. Danon, D. Wagner, E. Hideg, C. Gobel, I. Feussner, M. Nater & K. Apel. (2003) Rapid induction of distinct stress responses after the release of singlet oxygen in *Arabidopsis*. Plant Cell 15, 2320-2332.

Pecchioni N., P. Faccioli, H. ToubiaRahme, G. Vale & V. Terzi. (1996) Quantitative resistance to barley leaf stripe (*Pyrenophora graminea*) is dominated by one major locus. Theoretical and Applied Genetics 93, 97-101.

Pestsova E., J. Meinhard, A. Menze, U. Fischer, A. Windhovel & P. Westhoff. (2008) Transcript profiles uncover temporal and stress-induced changes of metabolic pathways in germinating sugar beet seeds. BMC Plant Biology 8, 122.

Ponquett R.T., M.T. Smith & G. Ross. (1992) Lipid autoxidation and seed ageing: Putative relationships between seed longevity and lipid stability. Seed Science Research 2, 51-54.

Price A.L., N.J. Patterson, R.M. Plenge, M.E. Weinblatt, N.A. Shadick & D. Reich. (2006) Principal components analysis corrects for stratification in genome-wide association studies. Nature Genetics 38, 904-909.

Priestley D.A., V.I. Cullinan & J. Wolfe. (1985) Differences in seed longevity at the species level. Plant Cell and Environment 8, 557-562.

Pritchard J.K., M. Stephens & P.J. Donnelly. (2000) Inference of population structure using multilocus genotype data. Genetics 155, 945-959.

Queval G., E. Issakidis-Bourguet, F.A. Hoeberichts, M. Vandorpe, B. Gakiere, H. Vanacker, M. Miginiac-Maslow, F. Van Breusegem & G. Noctor. (2007) Conditional oxidative stress responses in the *Arabidopsis* photorespiratory mutant *cat2* demonstrate that redox state is a key modulator of daylength-dependent gene expression, and define photoperiod as a crucial factor in the regulation of H_2O_2-induced cell death. Plant Journal 52, 640-657.

Rajjou L. & I. Debeaujon. (2008) Seed longevity: Survival and maintenance of high germination ability of dry seeds. Comptes Rendus Biologies 331, 796-805.

Rajjou L., Y. Lovigny, S.P.C. Groot, M. Belghaz, C. Job & D. Job. (2008) Proteome-wide characterization of seed aging in *Arabidopsis*: A comparison between artificial and natural aging protocols. Plant Physiology 148, 620-641.

Ratajczak E. & S. Pukacka. (2005) Decrease in beech (*Fagus sylvatica*) seed viability caused by temperature and humidity conditions as related to membrane damage and lipid composition. Acta Physiologiae Plantarum 27, 3-12.

Rauser W.E., R. Schupp & H. Rennenberg. (1991) Cysteine, γ-glutamylcysteine, and glutathione levels in maize seedlings: Distribution and translocation in normal and cadmium-exposed plants. Plant Physiology 97, 128-138.

Rhazi L., R. Cazalis, E. Lemelin & T. Aussenac. (2003) Changes in the glutathione thiol-disulfide status during wheat grain development. Plant Physiology and Biochemistry 41, 895-902.

Roberts E.H. (1973) Predicting the storage life of seeds. Seed Science and Technology 1, 499-514.

Romagosa I., F. Han, J.A. Clancy & S.E. Ullrich. (1999) Individual locus effects on dormancy during seed development and after ripening in barley. Crop Science 39, 74-79.

Rostoks N., S. Mudie, L. Cardle, J. Russell, L. Ramsay, A. Booth, J.T. Svensson, S.I. Wanamaker, H. Walia, E.M. Rodriguez, P.E. Hedley, H. Liu, J. Morris, T.J. Close, D.F. Marshall & R. Waugh. (2005) Genome-wide SNP discovery and linkage analysis in barley based on genes responsive to abiotic stress. Molecular Genetics and Genomics 274, 515-527.

Sattler S.E., L.U. Gilliland, M. Magallanes-Lundback, M. Pollard & D. DellaPenna. (2004) Vitamin E is essential for seed longevity, and for preventing lipid peroxidation during germination. Plant Cell 16, 1419-1432.

Schafer F.Q. & G.R. Buettner. (2001) Redox environment of the cell as viewed through the redox state of the glutathione disulfide/glutathione couple. Free Radical Biology and Medicine 30, 1191-1212.

Schopfer P., C. Plachy & G. Frahry. (2001) Release of reactive oxygen intermediates (superoxide radicals, hydrogen peroxide, and hydroxyl radicals) and peroxidase in germinating radish seeds controlled by light, gibberellin, and abscisic acid. Plant Physiology 125, 1591-1602.

Seal C.E., R. Zammit, P. Scott, T.J. Flowers & I. Kranner. (2010) Glutathione half-cell reduction potential and α-tocopherol as viability markers during the prolonged storage of *Suaeda maritima* seeds. Seed Science Research 20, 47-53.

Sen S. & D.J. Osborne. (1974) Germination of rye embryos following hydration-dehydration treatments - enhancement of protein and RNA-synthesis and earlier induction of DNA-replication. Journal of Experimental Botany 25, 1010-1019.

Sen S. & D.J. Osborne. (1977) Decline in ribonucleic-acid and protein-synthesis with loss of viability during early hours of imbibition of rye (*Secale cereale* L.) embryos. Biochemical Journal 166, 33-38.

Smirnoff N. (2010) Tocochromanols: Rancid lipids, seed longevity, and beyond. Proceedings of the National Academy of Sciences 107, 17857-17858.

Sosulski F.W. & G.I. Imafidon. (1990) Amino acid composition and nitrogen-to-protein conversion factors for animal and plant foods. Journal of Agricultural and Food Chemistry 38, 1351-1356.

Spaner D., L.P. Shugar, T.M. Choo, I. Falak, K.G. Briggs, W.G. Legge, D.E. Falk, S.E. Ullrich, N.A. Tinker, B.J. Steffenson & D.E. Mather. (1998) Mapping of disease resistance loci in barley on the basis of visual assessment of naturally occurring symptoms. Crop Science 38, 843-850.

Steffenson B.J., P.M. Hayes & A. Kleinhofs. (1996) Genetics of seedling and adult plant resistance to net blotch (*Pyrenophora teres f teres*) and spot blotch (*Cochliobolus sativus*) in barley. Theoretical and Applied Genetics 92, 552-558.

Steiner A.M. & P. Ruckenbauer. (1995) Germination of 110-year-old cereal and weed seeds, the Vienna Sample of 1877. Verification of effective ultra-dry storage at ambient temperature. Seed Science Research 5, 195-199.

Sun W.Q. & A.C. Leopold. (1995) The Maillard reaction and oxidative stress during aging of soybean seeds. Physiologia Plantarum 94, 94-104.

Swanson S.J., P.C. Bethke & R.L. Jones. (1998) Barley aleurone cells contain two types of vacuoles: Characterization of lytic organelles by use of fluorescent probes. Plant Cell 10, 685-698.

Tadokoro T. & S. Kanaya. (2009) Ribonuclease H: Molecular diversities, substrate binding domains, and catalytic mechanism of the prokaryotic enzymes. FEBS Journal 276, 1482-1493.

Tang W., T. Charles & R. Newton. (2005) Overexpression of the pepper transcription factor CaPF1 in transgenic virginia pine (*Pinus Virginiana* Mill.) confers multiple stress tolerance and enhances organ growth. Plant Molecular Biology 59, 603-617.

Varshney R.K., M. Baum, P. Guo, S. Grando, S. Ceccarelli & A. Graner. (2010) Features of SNP and SSR diversity in a set of ICARDA barley germplasm collection. Molecular Breeding 26, 229-242.

Walters C. (1998) Understanding the mechanisms and kinetics of seed aging. Seed Science Research 8, 223-244.

Walters C., L.M. Hill & L.J. Wheeler. (2005a) Dying while dry: Kinetics and mechanisms of deterioration in desiccated organisms. Integrative and Comparative Biology 45, 751-758.

Walters C., L.M. Wheeler & J.M. Grotenhuis. (2005b) Longevity of seeds stored in a genebank: Species characteristics. Seed Science Research 15, 1-20.

Wenzl P., J. Carling, D. Kudrna, D. Jaccoud, E. Huttner, A. Kleinhofs & A. Kilian. (2004) Diversity Arrays Technology (DArT) for whole-genome profiling of barley. Proceedings of the National Academy of Sciences of the United States of America 101, 9915-9920.

Wenzl P., H.B. Li, J. Carling, M.X. Zhou, H. Raman, E. Paul, P. Hearnden, C. Maier, L. Xia, V. Caig, J. Ovesna, M. Cakir, D. Poulsen, J.P. Wang, R. Raman, K.P. Smith, G.J.

Muehlbauer, K.J. Chalmers, A. Kleinhofs, E. Huttner & A. Kilian. (2006) A high-density consensus map of barley linking DArT markers to SSR, RFLP and STS loci and agricultural traits. BMC Genomics 7, 206

Wu L., Z. Zhang, H. Zhang, X.-C. Wang & R. Huang. (2008) Transcriptional modulation of ethylene response factor protein JERF3 in the oxidative stress response enhances tolerance of tobacco seedlings to salt, drought, and freezing. Plant Physiology 148, 1953-1963.

Xiong L.M., K.S. Schumaker & J.K. Zhu. (2002) Cell signaling during cold, drought, and salt stress. Plant Cell 14, S165-S183.

Yamaguchi-Shinozaki K. & K. Shinozaki. (2006) Transcriptional regulatiory networks in cellular responses and tolerance to dehydration and cold stresses. Annual Review of Plant Biology 57, 781-803.

Yu J., G. Pressoir, W.H. Briggs, I. Vroh Bi, M. Yamasaki, J.F. Doebley, M.D. McMullen, B.S. Gaut, D.M. Nielsen, J.B. Holland, S. Kresovich & E.S. Buckler. (2006) A unified mixed-model method for association mapping that accounts for multiple levels of relatedness. Nature Genetics 38, 203-208.

Zeng D.L., L.B. Guo, Y.B. Xu, K. Yasukumi, L.H. Zhu & Q. Qian. (2006) QTL analysis of seed storability in rice. Plant Breeding 125, 57-60.

Zhang L.Y., S. Marchand, N.A. Tinker & F. Belzile. (2009) Population structure and linkage disequilibrium in barley assessed by DArT markers. Theoretical and Applied Genetics 119, 43-52.

Zhu H., G. Briceno, R. Dovel, P.M. Hayes, B.H. Liu, C.T. Liu & S.E. Ullrich. (1999) Molecular breeding for grain yield in barley: An evaluation of QTL effects in a spring barley cross. Theoretical and Applied Genetics 98, 772-779.

Tab. S1: Detailed description of ICARDA subset used for association study. Single accessions No. correspond to consecutive numbering in Fig. 1 and are described by their history (landrace/cultivar), six- or two-rowed characteristic (Rows), country of origin, area of origin and Varshney-clusters according to Varshney *et al.* (2010) germplasm pool, main- and sub-clusters of phylogenetic tree (Fig. 1) and Q-groups of population structure (Q). AFR, Africa; APS, Arabian Peninsula; MEA, Middle East Asia; NEA, North East Asia; UNK, unknown origin.

No.	History	Rows	Origin (country)	Origin (area)	Varshney-cluster	Main-Cluster	Sub-Cluster	Q
1	landrace	six	Azerbaijan	NEA	la	A	I	Q3
2	landrace	six	Turkmenistan	NEA	la	A	I	Q3
3	landrace	six	Oman	APS	lc	A	I	Q3
4	landrace	six	Pakistan	NEA	la	A	I	Q3
5	landrace	six	Uzbekistan	NEA	la	A	I	Q3
6	landrace	two	Syria	MEA	lg	A	II	Q4
7	cultivar	two	Syria	MEA	ln	A	II	Q4
8	cultivar	two	ICARDA	UNK	lp	A	II	Q4
9	landrace	six	Egypt	AFR	lo	A	II	-
10	landrace	two	Turkmenistan	NEA	la	B	I	Q3
11	landrace	six	Iran	NEA	la	B	I	Q3
12	landrace	two	India	NEA	lw	B	I	Q3
13	landrace	six	Afghanistan	NEA	la	B	I	Q3
14	landrace	six	Pakistan	NEA	la	B	I	Q3
15	landrace	six	Pakistan	NEA	la	B	I	Q3
16	landrace	six	Iran	NEA	la	B	I	Q3
17	landrace	two	Afghanistan	NEA	la	B	I	Q3
18	landrace	two	Afghanistan	NEA	la	B	I	Q3
19	landrace	six	Turkmenistan	NEA	la	B	I	Q3
20	landrace	six	Afghanistan	NEA	la	B	I	Q3
21	landrace	six	China	NEA	la	B	I	Q3
22	landrace	six	China	NEA	la	B	I	Q3
23	landrace	six	China	NEA	la	B	I	Q3
24	landrace	six	Iran	NEA	la	B	I	Q3
25	landrace	six	Iran	NEA	la	B	I	Q3
26	landrace	six	Iran	NEA	la	B	I	Q3
27	landrace	six	Iran	NEA	la	B	I	Q3
28	landrace	six	Iran	NEA	la	B	I	Q3
29	landrace	six	Pakistan	NEA	la	B	I	Q3
30	landrace	six	Iran	NEA	la	B	I	Q3
31	landrace	two	India	NEA	la	B	I	Q3
32	landrace	six	Pakistan	NEA	la	B	I	Q3
33	landrace	six	Pakistan	NEA	la	B	I	Q3
34	landrace	six	Azerbaijan	NEA	la	B	I	Q3
35	landrace	six	Pakistan	NEA	la	B	I	Q3
36	landrace	six	Iran	NEA	la	B	I	Q3

No.	History	Rows	Origin (country)	Origin (area)	Varshney-cluster	Main-Cluster	Sub-Cluster	Q
37	landrace	six	Iran	NEA	Ia	B	I	Q3
38	landrace	two	Iran	NEA	Ia	B	I	Q3
39	landrace	six	Iran	NEA	Ia	B	I	Q3
40	landrace	two	Iran	NEA	Ia	B	I	Q3
41	landrace	six	Oman	APS	Ic	B	II	Q3
42	landrace	two	Oman	APS	Ic	B	II	Q3
43	landrace	six	Oman	APS	Ic	B	II	Q3
44	landrace	six	Egypt	AFR	Ia	B	III	Q3
45	landrace	two	Iran	NEA	Ia	B	III	Q3
46	cultivar		ICARDA	UNK	-	B	III	Q3
47	landrace	six	Pakistan	NEA	Ia	B	III	Q3
48	landrace	six	Iraq	MEA	Ia	B	IV	Q3
49	landrace	six	Jordan	MEA	Ia	B	IV	Q3
50	landrace	two	Syria	MEA	Ia	B	IV	Q3
51	landrace	six	Iraq	MEA	Ia	B	IV	Q3
52	landrace	six	Afghanistan	NEA	Ie	C	I	Q3
53	landrace	two	Afghanistan	NEA	Ie	C	I	Q3
54	landrace	six	Iran	NEA	Ie	C	I	Q3
55	landrace	six	Afghanistan	NEA	Ie	C	I	Q3
56	landrace	two	Iran	NEA	Ir	C	I	-
57	landrace	two	Ethiopia	AFR	Ib	C	II	-
58	cultivar	two	Eritrea	AFR	Ib	C	II	Q3
59	landrace	six	Ethiopia	AFR	Ib	C	II	-
60	cultivar	two	Eritrea	AFR	Ib	C	II	Q3
61	landrace	two	Saudi-Arabia	APS	Ic	C	III	Q3
62	landrace	two	Yemen	APS	Ic	C	III	Q3
63	landrace	two	Yemen	APS	Ic	C	III	-
64	landrace	two	Yemen	APS	Ic	C	III	-
65	cultivar	six	Ethiopia	AFR	Ih	C	III	
66	landrace	two	Saudi-Arabia	APS	Ic	C	III	-
67	landrace	two	Saudi-Arabia	APS	Ic	C	III	-
68	landrace	six	Iran	NEA	Id	C	IV	Q1
69	landrace	two	Turkey	NEA	Id	C	IV	Q1
70	landrace	two	Turkey	NEA	Id	C	IV	Q1
71	cultivar	two	Turkey	NEA	Iv	C	IV	Q4
72	cultivar	two	Turkey	NEA	Iv	C	IV	Q4
73	landrace	six	Syria	MEA	Ir	C	V	Q1
74	landrace	two	Syria	MEA	Is	C	V	Q1
75	landrace	two	Syria	MEA	-	C	V	Q1
76	landrace	two	Syria	MEA	Is	C	V	Q1
77	landrace	six	Jordan	MEA	Is	C	V	Q1
78	landrace	two	Syria	MEA	Is	C	V	Q1
79	landrace	six	Syria	MEA	Is	C	V	Q1
80	landrace	two	Jordan	MEA	Is	C	V	Q1

No.	History	Rows	Origin (country)	Origin (area)	Varshney-cluster	Main-Cluster	Sub-Cluster	Q
81	cultivar	two	ICARDA	UNK	Is	C	V	Q1
82	landrace	two	Jordan	MEA	Is	C	V	Q1
83	landrace	six	Syria	MEA	It	C	V	Q1
84	cultivar	two	ICARDA	UNK	It	C	V	Q1
85	cultivar	two	ICARDA	UNK	It	C	V	Q1
86	cultivar	two	ICARDA	UNK	It	C	V	Q1
87	cultivar	two	ICARDA	UNK	Is	C	V	Q1
88	landrace	six	Syria	MEA	Is	C	V	Q1
89	cultivar	two	ICARDA	UNK	-	C	V	Q1
90	cultivar	two	ICARDA	UNK	-	C	V	-
91	cultivar	two	Libya	AFR	Ir	C	-	Q1
92	landrace	six	Algeria	AFR	If	D	I	Q4
93	landrace	six	Tunisia	AFR	If	D	I	Q4
94	cultivar	six	France	EUR	If	D	I	Q4
95	cultivar	six	Russia	EUR	If	D	I	Q4
96	landrace	two	Jordan	MEA	If	D	I	Q4
97	landrace	six	Uzbekistan	NEA	Ig	D	I	Q4
98	landrace	two	Syria	MEA	Ig	D	I	Q4
99	landrace	two	Syria	MEA	Ig	D	I	Q4
100	landrace	six	Turkmenistan	NEA	Ig	D	I	Q4
101	cultivar	six	ICARDA	UNK	-	D	I	Q4
102	cultivar	six	ICARDA	UNK	Ih	D	I	Q4
103	cultivar	six	Tunisia	AFR	Ih	D	I	Q4
104	cultivar	six	ICARDA	UNK	-	D	I	-
105	landrace	six	Turkey	NEA	Ig	D	I	Q4
106	landrace	six	Azerbaijan	NEA	Ig	D	II	Q4
107	landrace	six	Egypt	AFR	-	D	II	-
108	landrace	six	Egypt	AFR	Io	D	II	Q1
109	cultivar	two	Syria	MEA	In	D	II	Q4
110	cultivar	two	ICARDA	UNK	Ip	D	II	Q4
111	cultivar	two	ICARDA	UNK	Ip	D	II	Q4
112	cultivar	two	ICARDA	UNK	Ip	D	II	Q4
113	cultivar	two	ICARDA	UNK	Ip	D	II	Q4
114	cultivar	two	ICARDA	UNK	In	D	II	Q4
115	cultivar	two	Lebanon	MEA	In	D	II	Q4
116	cultivar	two	ICARDA	UNK	In	D	II	Q4
117	cultivar	two	ICARDA	UNK	Ip	D	II	Q4
118	landrace	two	Egypt	AFR	-	D	II	Q4
119	landrace	two	Saudi-Arabia	APS	Iq	D	II	Q4
120	landrace	six	Oman	APS	Iq	D	II	Q4
121	landrace	two	Oman	APS	Iq	D	II	Q4
122	cultivar	two	Australia	AUS	Ip	D	II	Q4
123	cultivar	two	ICARDA	UNK	Ip	D	II	Q4
124	cultivar	two	Australia	AUS	Ip	D	II	Q4

No.	History	Rows	Origin (country)	Origin (area)	Varshney-cluster	Main-Cluster	Sub-Cluster	Q
125	cultivar	two	ICARDA	UNK	Ip	D	II	Q4
126	cultivar	two	ICARDA	UNK	Ip	D	II	Q4
127	landrace	six	Libya	AFR	Ii	E	I	Q2
128	landrace	six	Tunisia	AFR	Ii	E	I	Q2
129	landrace	six	Algeria	AFR	Ii	E	I	Q2
130	landrace	six	Algeria	AFR	Ii	E	I	Q2
131	landrace	six	Algeria	AFR	Ii	E	I	Q2
132	landrace	six	Algeria	AFR	Ii	E	I	Q2
133	landrace	six	Iran	NEA	Ii	E	I	Q2
134	cultivar	six	Algeria	AFR	Ii	E	I	Q2
135	landrace	six	Algeria	AFR	Ii	E	I	Q2
136	landrace	six	Tunisia	AFR	Ii	E	I	Q2
137	landrace	six	Egypt	AFR	Ii	E	I	Q2
138	cultivar	six	Egypt	AFR	Ii	E	I	Q2
139	cultivar	six	Egypt	AFR	Ii	E	I	Q2
140	landrace	six	Egypt	AFR	Il	E	I	Q2
141	landrace	six	Libya	AFR	Im	E	I	Q2
142	landrace	six	Algeria	AFR	Ii	E	I	Q2
143	landrace	six	Tunisia	AFR	Ii	E	I	Q2
144	cultivar	six	Tunisia	AFR	Ii	E	I	Q2
145	landrace	six	Libya	AFR	Ii	E	II	Q2
146	cultivar	six	Jordan	MEA	Ik	E	II	Q2
147	cultivar	six	Libya	AFR	Il	E	II	Q2
148	cultivar	six	Cyprus	NEA	Ik	E	II	-
149	cultivar	six	Greece	EUR	Ik	E	II	-
150	cultivar	six	Syria	MEA	I-	E	II	-
151	landrace	six	Morocco	AFR	Im	E	III	Q2
152	landrace	six	Morocco	AFR	Ii	E	III	Q2
153	landrace	six	Morocco	AFR	Ii	F	III	Q2
154	landrace	six	Morocco	AFR	Ii	E	III	Q2
155	landrace	six	Egypt	AFR	Ii	E	III	Q2
156	landrace	six	Egypt	AFR	Ii	E	III	Q2
157	landrace	six	Libya	AFR	Ii	E	III	Q2
158	landrace	six	Libya	AFR	Ii	E	III	Q2
159	landrace	six	Libya	AFR	Ii	E	III	Q2
160	landrace	six	Libya	AFR	Ii	E	III	Q2
161	landrace	six	Libya	AFR	Ii	E	III	Q2
162	landrace	two	Syria	MEA	Ij	E	IV	Q2
163	landrace	six	Oman	APS	Ij	E	IV	Q2
164	cultivar	six	Egypt	AFR	Ij	E	IV	Q2
165	landrace	six	Libya	AFR	Ij	E	IV	Q2
166	cultivar	six	ICARDA	UNK	Ij	E	IV	Q2
167	cultivar	six	Iraq	MEA	Ij	E	IV	Q2
168	cultivar	six	Morocco	AFR	Ij	E	IV	Q2

No.	History	Rows	Origin (country)	Origin (area)	Varshney-cluster	Main-Cluster	Sub-Cluster	Q
169	cultivar	six	ICARDA	UNK	lj	E	IV	Q2
170	cultivar	six	ICARDA	UNK	lj	E	IV	Q2
171	landrace	two	Tajikistan	NEA	lv	E	IV	-
172	landrace	two	Tajikistan	NEA	lv	E	IV	-
173	cultivar	six	Algeria	AFR	li	E	V	Q2
174	landrace	six	Egypt	AFR	lm	E	V	-
175	cultivar	six	Libya	AFR	li	E	V	-

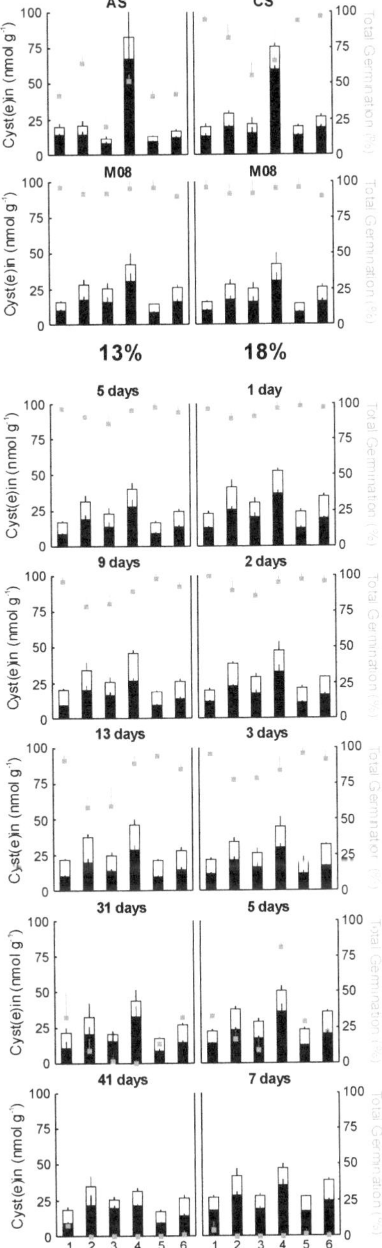

Fig. S1: Relationship between seed viability and cyst(e)ine for six genotypes in response to seed storage and ageing. Seeds from six selected genotypes of the Gatersleben barley collection were tested after 10 to 14 years cold storage (CS at 0°C) and, ambient storage (AS at 20°C) for the same period. In 2008, seeds were multiplied from CS material stored at 0°C and subjected to artificial ageing. M08, non-aged control; CD, controlled deterioration at 45°C and 13 or 18% seed MC for 5, 9, 13, 31 and 41 d (for 13% seed MC) and 1, 2, 3, 5 or 7 d (for 18% seed MC). Total germination is shown in blue, black bars show disulphide cystine and white bars show thiol cysteine. Data are mean ±SD of five independent biological replicates.

Fig. S2: Relationship between seed viability and γ-glutamylcyst(e)ine for six genotypes in response to seed storage and ageing. Seeds from six selected genotypes of the Gatersleben barley collection were tested after 10 to 14 years cold storage (CS at 0°C) and, ambient storage (AS at 20°C) for the same period. In 2008, seeds were multiplied from CS material stored at 0°C and subjected to artificial ageing. M08, non-aged control; CD, controlled deterioration at 45°C and 13 or 18% seed MC for 5, 9, 13, 31 and 41 d (for 13% seed MC) and 1, 2, 3, 5 or 7 d (for 18% seed MC). Total germination is shown in blue, black bars show disulphide γ-glutamylcystine and white bars show thiol γ-glutamylcysteine. Data are mean ±SD of five independent biological replicates.

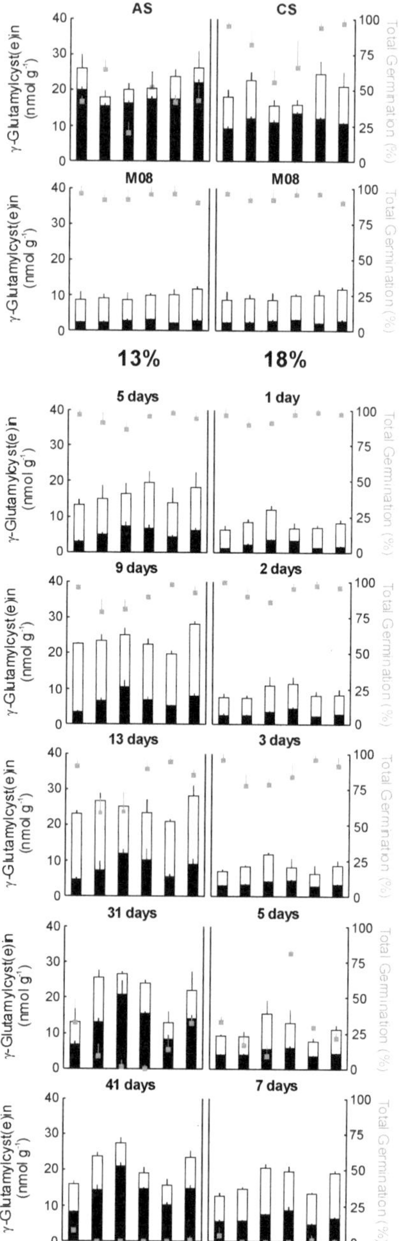

Fig. S3: Relationship between seed viability and cyst(e)inylglycine for six genotypes in response to seed storage and ageing. Seeds from six selected genotypes of the Gatersleben barley collection were tested after 10 to 14 years cold storage (CS at 0°C) and, ambient storage (AS at 20°C) for the same period. In 2008, seeds were multiplied from CS material stored at 0°C and subjected to artificial ageing. M08, non-aged control; CD, controlled deterioration at 45°C and 13 or 18% seed MC for 5, 9, 13, 31 and 41 d (for 13% seed MC) and 1, 2, 3, 5 or 7 d (for 18% seed MC). Total germination is shown in blue, black bars show disulphide cystinylglycine and white bars show thiol cysteinylglycine. Data are mean ±SD of five independent biological replicates.

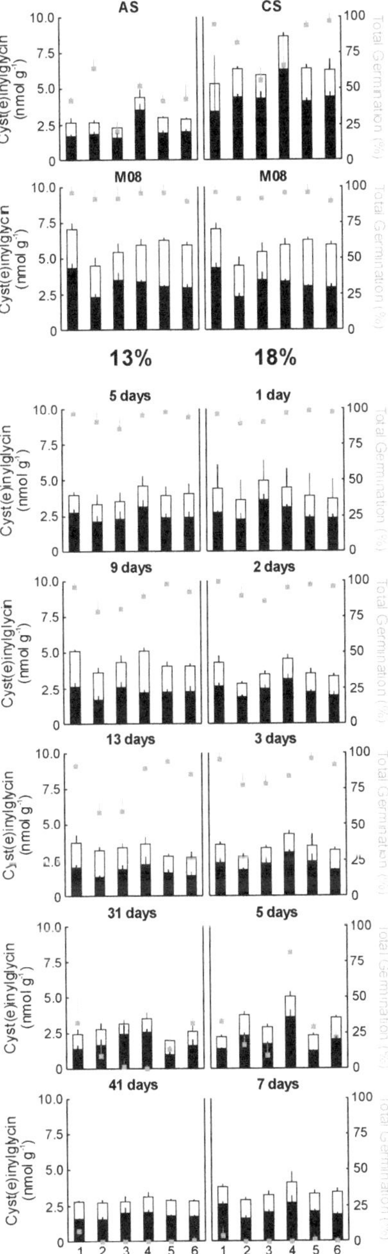

5   Epilog

During the last millennia farmers have grown seeds for the purpose of food and feed production and used the best material from the last harvest for the next season. Therefore, seed longevity had never been considered as an important trait until seedbanks appeared and the issue how long transgenic seeds might survive in soil arose.

5.1 Genotypic differences in seed longevity: analysis of genebank data

Only few studies were conducted to investigate seed life-span in crops. Among them are investigations of long-term survival of seeds under ambient conditions (Haferkamp *et al.* 1953, Aufhammer & Simon 1957, Priestley *et al.* 1985, Nagel & Börner 2010). These studies are important for farmers in temperate zones who mainly use this storage procedure and for vegetable seed producer (Roos & Davidson 1992) who pay high production expenses. However, due to the complexity and duration of experiments very little is known about survival of crop seeds in soil (Kivilaan & Bandurski 1981, Rauber 1987), in ultra-dry storage (Steiner & Ruckenbauer 1995), in cold storage (Walters *et al.* 2005a) and in cryopreservation (Walters *et al.* 2004).

For examination of molecular and biochemical processes of seed deterioration the experimental procedure was expedited up by 'Accelerated Ageing' and 'Controlled Deterioration' which are included in the ISTA (International Seed Testing Association) vigour test methods (Hampton & TeKrony 1995, ISTA 2008). Both methods promote the reduction of seed vigour by increasing the storage temperature and moisture content of seeds and are good indicators for field emergence and storability (Hampton & TeKrony 1995). On basis of proteomic analysis Rajjou *et al.* (2008) could demonstrate a similarity between controlled deterioration procedures and natural ageing processes. Already in 1960, Roberts (1960) found a mathematical relationship between the decrease of seed vigour and the storage conditions particularly the seed moisture content and temperature. The so called viability equation is applicable to orthodox seeds in the range of −20 to 90°C and 5 to 25% seed moisture content and depends on various factors and constants (Ellis & Roberts 1980b).

$$v = K_i - \frac{p}{10^{K_E - C_W \log m - C_H t - C_Q t^2}}$$

v: final viability (in %) after p days of storage
p: storage time (in years or days)
m: seed moisture content (smc in %) in equilibrium
t: storage temperature (°C)
K_i: initial viability of the seed lot at p = 0 days (seedlot constant)
C_H and C_Q: species-specific temperature constants
K_E and C_W: species-specific moisture content constants

The viability equation requires species-specific constants which are determined by a series of experiments using several combinations of temperature and moisture content (Hay *et al.* 2003). These constants are publicly available for about 66 mainly crop species (KEW 2008).

It was assumed that genotypic effects do not have any influence as demonstrated in *Arabidopsis* (Hay *et al.* 2003). However, Nagel & Börner (2010), Kochanek *et al.* (2009), Revilla *et al.* (2006, 2009), Walters *et al.* (2005a), Tesnier *et al.* (2002) and also Roberts and Ellis (1977) considered a genotypically controlled variation in longevity within species. In this case the equation would work only for few genotypes or the constants need to be adapted.

The presented germination data (due to a precise distinction with total germination also determined as vigour) of six different genebank species in paper 1 gave first evidence for genetic control of the trait longevity after long-term storage. All genebank accessions of the same species had the same harvest year including similar growing, harvest, cleaning and storage conditions in common. The separation of initially high germinations is assumed to be genetic.

Nevertheless, the utilisation of genebank information and accessions is inevitably associated with several issues because the handling of the material was not designed for long-term experiments. Fig. 7 shows 150 barley accessions harvested in three years and stored at 0°C for about 35 years. The inserted black line is the estimated viability curve of 50 accessions using probit analyses which enable comparisons between the harvest years (Nagel *et al.* 2011, unpublished data). The curves show a slope increase from 1974 to 1976 which is linked with the doubling of the half-viability period P50. Dependent on the year, the analyses of germination tests revealed different results which are due to a variation of factors we were not able to control in the past and that are explained as follows:

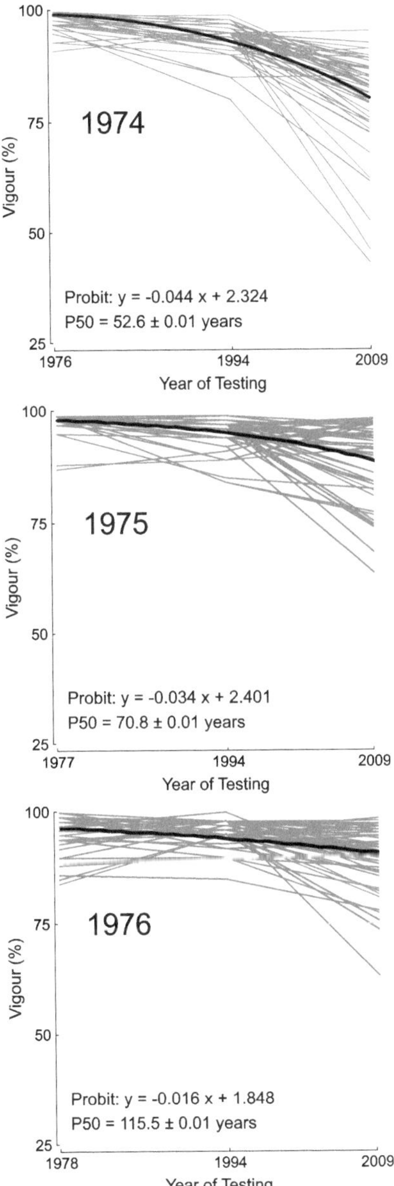

Fig. 1: Vigour test results (ISTA germination) of 150 barley accessions harvested in 1974, 1975 and 1976 and stored at 0°C until 2009 (Nagel *et al.* 2011, unpublished data). Probit shows estimated linear equation of 50 accessions per year. P50, half-viability period; Black line, estimated probit curve transformed from the linear probit equation (Probit) into percentages.

1) The genotypes of each year are different dependent on the necessity of regeneration after long-term storage. Genotype effects are hard to discover due to missing annual repetitions. Furthermore different accessions produce phenotypes with varying long-living behaviour. Therefore a bias distortion cannot be excluded.

2) The field plots changed every year because of a required crop rotation. A randomisation was not applied and the nutrient supply was unknown.

3) The weather conditions in June, July and August had become warmer from 1974 to 1976 (average temperature at 2m: 1974: $16.0 \pm 1.4°C$; 1975: $18.1 \pm 2.4°C$; 1976: $18.0 \pm 1.8°C$) and drier (average rainfall: 1974: 44.1 ± 13.5 mm; 1975: 36.4 ± 42.2 mm; 1976: 27.0 ± 14.7 mm). Further external factors as insect and pathogen attacks are partly connected with weather and influence the plant development and the final seed quality.

4) The history of the material in storage is not fully documented. The German federal *ex situ* genebank is a living collection which supplies seeds for scientists, companies and private users. Therefore a part of the material is frequently used. Unfortunately, the initial seed moisture content, the number of withdrawals and conditions during these procedures cannot be unveiled.

5) Germination test conditions performed in the past are not totally described. Additionally the limited seed supply of genebank accessions enables only tests of small sub-samples. Also the presented final germination data are performed with 50 seeds in 3 replications which produced in some cases a maximum standard deviation of $\pm15.3\%$.

Most of the described factors influence the initial seed quality (K_i) which is a main component of the viability equation and has a significant impact on seed storage behaviour (Probert *et al.* 2007). To demonstrate the effect of K_i and the seed moisture content, the minimum and maximum initial germination and different seed moisture contents (with respect to the final measurements of Fig. 8 of the three harvest years) were included in the viability equation and are represented in Tab. 2.

Tab. 1: Application of viability equation (Ellis & Roberts 1980b) using initial data of 150 previously described barley accessions of the harvest year 1974, 1975, 1976 retested in 2009. According to Ellis & Roberts (1980a) and Dickie *et al.* (1990) barley constant are $K_E = 9.144$, $C_W = 5.342$, $C_H = 0.0329$, $C_Q = 0.000478$, Initial vigour = K_i (%), smc = seed moisture content.

Harvest year	1974		1975		1976	
	Min	Max	Min	Max	Min	Max
Storage periode (p in years)	35		34		33	
Initial vigour (K_i in %)	91.0	100	89.0	100	84.0	100
Final vigour at **5% smc**	90.2	100	88.1	100	82.8	100
Final vigour at **7% smc**	85.1	100	82.5	100	76,2	100
Final vigour at **9% smc**	57.6	100	54,4	100	46,5	100
Final vigour at **11% smc**	2.2	91.9	2.1	93.3	1.5	94.4
Final vigour at **13% smc**	0	0	0	0	0	0
Final vigour at **15% smc**	0	0	0	0	0	0

The final seed moisture contents of the 150 barley accessions were determined on basis of water activity which enables a non destructive measurement. Sorption isotherms of barley (Ertugay & Certel 2000) reveal finally the water content at 22.8 ± 1°C. The tested material showed a water activity in the range of 0.15 to 0.73 which is equal to 5.0 and 14.8% smc. If these moisture contents had been the initial ones they would strongly affect the final germination as shown in Tab. 2. Assuming additionally a K_i of 84.0% (1976) the germination could drop up to 0% (Tab. 2). However, the lack of knowledge about the storage history of the examined accessions provides several explanations possibilities for the observed germination separation. Nevertheless, a correlation between the final seed moisture content and final germination was not detected. Furthermore there was no relation between the initial and the final germination. This observation was also supported by results for sorghum (*Sorghum bicolor* L.), wheat, oilseed rape (*Brassica napus* L.), rye and flax (*Linum usitatissimum* L.) (first paper) and confirms the assumption of a genetic component.

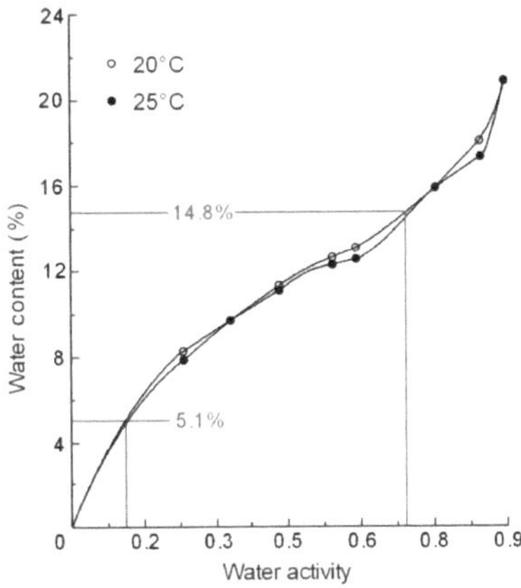

Fig. 2: Barley sorption isotherms at 20 and 25°C (Ertugay & Certel 2000). Red lines indicate minimum and maximum measured water activity of 150 barley accessions.

In general, genetic control of seed longevity cannot be excluded. Nearly 100 genes have been identified to underlie natural variation for plant development and growth (Alonso-Blanco *et al.* 2009) and are exemplified by flowering time (Stracke *et al.* 2009, Cockram *et al.* 2011), disease resistance (Williams 2003) and a wide range of other agricultural traits (Wenzl *et al.* 2006) in barley. The genetic control of various traits leads also to a variation in seed survival which is demonstrated in *Arabidopsis*. Here, structural and pigmentation changes in testa result in divers longevities (Debeaujon *et al.* 2000). The abundance of seed storage proteins and optimum developmental processes improved the storage behaviour (Sugliani *et al.* 2009). Additionally the mapping populations mentioned in the second and third paper of this thesis, experiments in rice (Miura *et al.* 2002, Zeng *et al.* 2006), lettuce (*Lactuca sativa* L.) (Schwember & Bradford 2010) and *Arabidopsis* (Bentsink *et al.* 2000, Clerkx *et al.* 2004) gave strong evidence for the genetic background of seed longevity variations.

Considering the genetic impact and various factors influencing K_i, the viability equation is hardly useful for the prediction of storage intervals and final germination data in genebanks. Further limitations regarding temperatures and moisture content (Dickie *et al.* 1990) have provoked new research. On basis of the molecular mobility which is a key factor for the storage stability a linear relationship between ageing rates of germplasm and the logarithm of rotational motions in the cytoplasm was established. The prediction of vigour loss and longevity was enabled only by measuring rotational correlation times at low temperature (Buitink *et al.* 2000). Unfortunately this procedure requires elaborated equipment and does not enable a quick estimation.

Walters *et al.* (2005a) continued the development of viability equations based on parameters which can be applied to different storage conditions. A special form of the Avrami equation (Avrami 1941, Williams *et al.* 1993) which describes reactions based on visco-elastic properties in relation with glassy relaxation was used to study ageing kinetics in seeds. The extrapolated curve shapes had remarkable similarities with deterioration curves (Walters 1998).

Summarising, genotypic influences on seed longevity have been widely proved by a number of authors. Unique genebank data of long-term stored material could also confirm these findings. However, due to the lack of knowledge of the history of the different accessions the assumption is equipped with several issues. Nevertheless, the difference in germination after several years of storage endangers unique accessions which need to be monitored more frequently. The adaptation of viability equations to genotypes or new methods of seed storability prediction integrating biochemical and physical knowledge could help to estimate regeneration periods and to reduce the risk of losing valuable accessions.

5.2 Genetics of seed longevity

Since centuries it has been a feverish desire of mankind to prolong lifespan by an elixir and to unveil the secret of ageing. Even in the 21st century scientists are far away from unravelling these big secrets and the enormous variation in longevity among species points towards genetic factors affected by multiple loci (Vijg & Suh 2005).

The longevity of seed has not been important up to the appearance of genebanks and the necessity to safe unique genotypes in long-term storage. However, revealing the secret of ageing is associated with experimental challenges which are comparable over a wide range of species. Monaghan *et al.* (2008) defined ageing as an irreversible accumulation of damage that leads to a time-dependent loss of fitness and eventual death. It begins after the organism attains its maximum reproductive competence (Vijg & Suh 2005) and proceeds with strong environmental influences (Monaghan *et al.* 2008). Unfortunately the loss of fitness cannot be easily caught by biomarkers because its appearance is extremely complex (Vijg & Suh 2005). In case of seeds it is even more sophisticated because neither changes in phenotypes occur nor does the death become visible. The only way identifying the viability status of seeds is to perform germination tests and thus to destroy the seed. Currently there is no other option to overcome this problem.

However, to understand the genetic basis of complex traits such as longevity scientists focused traditionally on quantitative genetics. Two distinct methods are contemporary used: the quantitative trait locus (QTL) or linkage mapping and the association or linkage disequilibrium mapping including marker-trait associations (MTAs) (Hall *et al.* 2010). In the present studies both methods were applied to examine the genetic background of seed longevity in barley and are characterised with different advantages and disadvantages.

In QTL mapping, an early generation cross is used to dissect quantitative variation that separates the individuals of the parental generation (Hall *et al.* 2010). This method has been proven to be a powerful and accurate way in identifying many genome regions (Price 2006). Nevertheless, it has still some limitations due to the use of biparental populations. The two parental lines will only segregate for alleles that differ between each other. Consequently more than one population is required to present the genetic architecture of a trait (Holland 2007). To study the genetic background of seed longevity, three different populations were applied in the second paper: the Steptoe-Morex (SxM), the Oregon Wolfe Barley (OWB) and the W766 population. The SxM has been developed to solve practical breeding problems and derived from a cross between a high yielding six-rowed feed-type barley and a six-rowed cultivar used as the American malting industry standard (Kleinhofs *et al.* 1993). The OWB population consist of spring barley lines developed by Costa *et al.* (2001) on basis of morphological variants (e.g. flowering time, dwarfing). The winter barley population W766 is a cross between a two-rowed cultivar and a short-stemmed, dense-eared two-rowed Japanese landrace (Buck-Sorlin 2002).

Phenotyping the three mapping populations revealed in total four QTLs of the trait germination after artificial ageing expressed as seed longevity. Only one QTL was found at the same position in the OWB and W766 population. Three further loci were found either in the OWB or in the SxM population demonstrating the restricted allelic variation of only two parents used in the populations.

Unfortunately, QTL mapping suffers also from the numbers of recombination events per chromosome which are usually small and limit the resolution of the genetic map. A typical QTL identified from a population of a few hundred lines can encompass a few to tens of centiMorgans (cM), which might correspond to genomic regions of several megabases including hundreds or even thousands of genes (Price 2006, Hall *et al.* 2010). This was also the case in the OWB, SxM and W766 population. The QTLs of these studies covered at least 10 to 30 cM. Additionally the generation of mapping populations is very time consuming or in case of forest trees sometimes even impossible because of long generation intervals (Hall *et al.* 2010).

To overcome the limitations of QTL mapping, association mapping provides an alternate route by using a broader spectrum of germplasm (Holland 2007). In vertebrates where controlled crosses can be expensive or for humans even impossible this method gives new opportunities to search for genotype-phenotype correlations (Myles *et al.* 2009). In general they result in higher resolutions because they involve natural populations which reflect historical combination events (Nordborg & Weigel 2008). However, due to the limited extend of linkage disequilibrium (LD), which is defined as a non-random association of alleles at separate loci located on the same chromosome (Mackay & Powell 2007), a higher number of genetic markers is needed to ensure adequate power to detect a linkage between a marker and a causal locus (Hall *et al.* 2010). Especially the LD of barley, a self-pollinating species, is predicted to be extensive but can also decay within a short range which is caused by recombination events (Waugh *et al.* 2009).

Further critics arose on the presence of a population structure which defines the genetic architecture of a trait in a sample of individuals. This structure depends on the fact how the material is assembled and can be a confounding variable especially for traits that are important in adaptation to the local environment (Nordborg & Weigel 2008). For example a phenotypic trait could correlate with the underlying population structure at neutral, unaffected loci and show an inflated number of positive associations (Hall *et al.* 2010).

However, to benefit from the association mapping methods we used in the third paper a collection (henceforth termed population) selected by the International Center for Agricultural Research in the Dry Area (ICARDA) that was designed for the exploration of plant adaptation to climate change in particular drought stress (Varshney *et al.* 2010). The applied accessions consisted of 53 cultivars and 122 landraces derived from 30 countries of mainly dry areas. The population structure was not clearly clustered but gave a hint of accessions origin and their usage. The plant material was genotyped with 703 DArT markers resulting in an average distance of 1.6 cM. Unfortunately also gaps with marker distances of at least 10 cM have been detected in 21 positions. To overcome the risk of false-positives the analyses was performed with two different models that use information of population structure based on Principle Component Analyses (Price *et al.* 2006) and the relative kinship of the individuals (Yu *et al.* 2006). An MTA was assigned if both methods appeared significant for one seed longevity trait and additionally two of in total six traits fulfilled this requirement. Finally 105 MTAs at 32 loci were detected for vigour and 98 MTAs at 28 loci for total germination. Vigour and total germination differ in the amount of germinated seed which are devided in normal and abnormal seedlings in case of vigour.

Taking only total germination in account and using the consensus map of Wenzl *et al.* (2006) QTLs of OWB, SxM and W766 population were co-located with MTAs on chromosomes 2H, 5H, 7H and confirmed these previous results. Additionally seed longevity QTLs of rice detected by Miura *et al.*(2002), Zeng *et al.* (2006) and Xue *et al.* (2008) can be roughly located also in comparable regions of chromosomes 5H and 2H (Fig. 9).

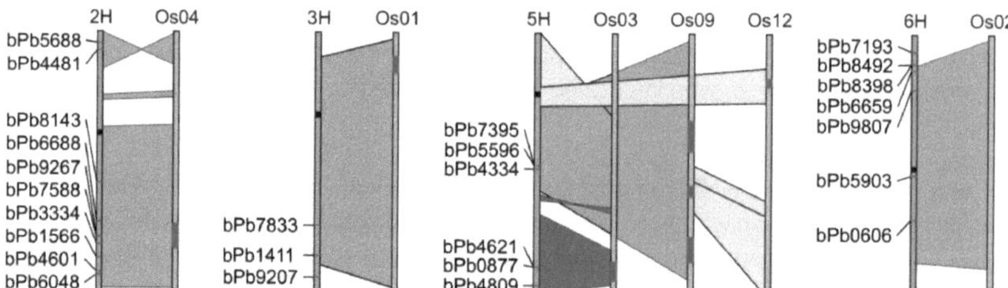

Fig. 3: Comparison of longevity loci in barley (H) and rice (Os) on basis of total germination of the present association mapping study and germination data of Miura *et al.* (2002), Zeng *et al.* (2006) and Xue *et al.* (2008) using colinearity information between the barley consensus map and rice chromosomes given by Stein *et al.* (2007). Red lines indicate MTAs of the present study, red squares correspond with QTLs for seed longevity in rice, black square gives position of centromere in barley.

Seed longevity loci were also found in soybean (*Glycine max* L.) (Singh *et al.* 2008), maize (Revilla *et al.* 2009), *Arabidopsis* (Clerkx *et al.* 2004) and lettuce (Schwember & Bradford 2010). Due to limited synteny of barley to these other species and the current unavailability of synteny maps, it is only possible to speculate about similar positions of longevity loci and an overall mechanism of seed ageing.

However, seed longevity studies are generally conducted for gaining knowledge about the mechanism of seed deterioration in storage which can be used for seed storage improvements in the future. Unfortunately, storage conditions can be widely defined. In temperate zones the storage of orthodox seeds is usually carried out on farms under ambient conditions (~20°C, ~50 RH), in seedbanks under cold storage conditions (-18°C, ~5% smc), in soil seed banks under strong alterable conditions concerning temperature and moisture content in winter and summer and in many transition zones. Depending on the species and the conditions seeds can survive periods between years and centuries (Priestley *et al.* 1985, Walters *et al.* 2005a, Nagel & Börner 2010) as it has already been mentioned. After real-time maintenance, meaning storage periods of tens of years, genetic backgrounds have not been investigated yet. Therefore the deteriorative effects of high seed moisture contents and high temperature are usually applied to save time for genetic studies.

The ISTA defined two procedures which are the basic approaches for most seed longevity assays: the accelerated ageing test and the controlled deterioration test (Hampton & TeKrony 1995). The two methods distinguish themselves in the availability of water. The accelerated ageing is performed in the presence of free available water. Seeds reach moisture equilibrium during high temperature treatment (30 to 60°C). In contrast a defined amount of water is added to achieve specific seed moistures in controlled deterioration test. Seeds can equilibrate to this moisture content before the temperature is increased (Hampton & TeKrony 1995). Both treatments applied for several hours until days and months decrease seed germination which is assumed to mimic natural ageing (Delouche & Baskin 1973, Priestley *et al.* 1980, Bentsink *et al.* 2000). Additionally a proteome analysis revealed that controlled deterioration could also show similar pattern to long-term stored seed with respect to oxidation of seed proteome and identical protein carbonylation (Rajjou *et al.* 2008).

In fact the artificial ageing procedure might produce similar results compared to some long-term storage methods but the increase of moisture content will change the hydration level which results in an initiation of metabolic processes (Walters *et al.* 2005b). The third paper (Fig. 6) shows in detail that during the first days of ageing the vigour increases which indicates an effect that is well known from priming. This technique involves the controlled uptake of water followed by drying and initiates the activation of enzymes that are necessary for breakdown of storage reserve during germination. Furthermore also the glutathione assay results in significant different amounts of total glutathione between different artificial ageing and long-term storage methods.

Nevertheless, to reduce the experimental time from years or decades to days and months we were also forced to use different ageing procedures. In the second paper both of the described methods were applied: the accelerated ageing test (43°C, 100% RH, 72h) and the controlled deterioration test (44°C, 18% smc, 72h). In all three mapping populations both methods revealed the same QTLs. However, this result has not been proved in all species as shown by Schwember & Bradford (2010) and Nagel *et al.* (2011). Schwember and Bradford (2010) could only prove poor correlations between controlled deterioration and storage at 9°C and 30% RH of lettuce seed. In case of an oilseed rape population, the temperature reduction of 1°C resulted in complete different QTL (Fig. 10). Despite the fact that high moisture levels (43°C, 100% RH, 72h) and low moisture levels (45°C, 60% RH, 15d) were used for the artificial ageing of the present association mapping study (third paper) only few ageing specific loci were found due to our experimental design. The aim of this was to find an overall mechanism for ageing. Therefore the ageing variants were mixed to form an MTA.

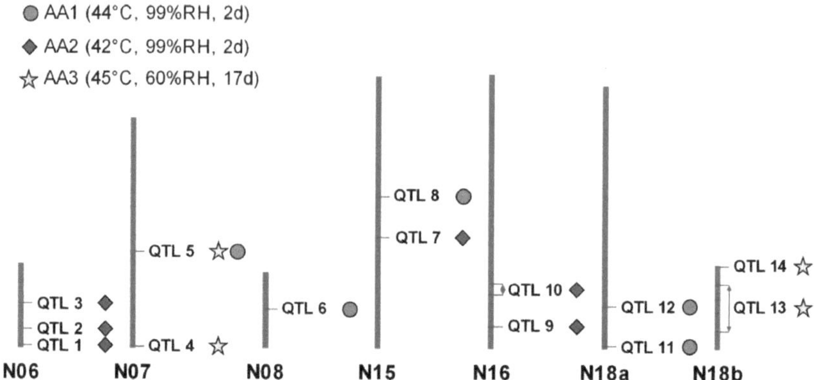

Fig. 4: Seed longevity QTLs of oilseed rape population YE2-DH with respect to different artificial ageing (AA) conditions. Figure is adapted from Nagel *et al.* (2011).

Regarding the diversity of storage methods there is still the assumption that different mechanisms are responsible for seed death. Therefore seeds might develop different strategies to protect themselves against deterioration (Walters 1998, Walters *et al.* 2005b). Some of the presented strategies for unlocking the genetic and biochemical secrets of seed longevity might correlate with these defence strategies. However, this can be only known for certainty if the material is stored under these corresponding conditions.

5.3 Biochemical aspects of seed ageing

Biochemical processes occurring in plants and finally in seeds are controlled by the genetic background of the plant material. Using quantitative genetics, our study of seed longevity revealed a variety of loci which are assumed to be involved in the long-term survival of seeds. Fig. 11 provides a generalised overview of the possible functions of marker sequences and the co-located QTLs detected in the second and third paper. More than two third of QTLs derived from different studies can be classified into agronomic characteristics that include traits such as yield, thousand kernel weight (TKW), plant height and days to maturity. The remaining one third is divided up on QTLs based on germination, malting properties as for example soluble protein, α-amylase activity and response to diseases. In case of assumed annotated functions, 61% are related to stress response, resistance or are involved in signalling, one third is linked with plant development and transport and 11% are related to transcription and translation processes.

In this context one question arises: did we find the majority of QTLs and annotated functions because of their main responsibility for seed longevity or because more studies were performed due to the economic relevance of agronomic factors and to overcome pathogens and abiotic stress by resistance pathways? One part of the answer could be that abiotic and biotic stresses negatively influence the survival, biomass production and crop yield (Agarwal *et al.* 2006) which is also associated with the initial quality of seed expressed as K_i in the seed viability equation. Therefore, the trait seed longevity or the detected loci seemed to be strongly involved in the whole plant development which also includes the dependency on environmental conditions as climatic factors or the appearance of plant pathogens. The arrangement of the association mapping experiment on two field plots allowed to compare different environments of the same year. The initial vigour was significantly reduced from the plot with limited nutrient and organic matter supply. Additionally in 18 out of 32 vigour loci the initial vigour was involved. This result supports the assumption that loci influence seed longevity before maturation.

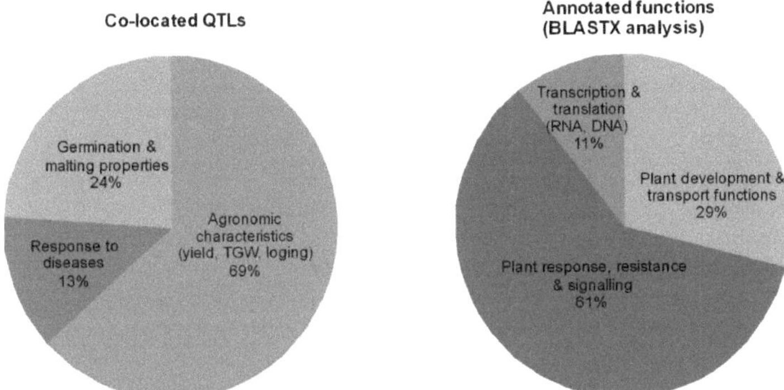

Fig. 5: **Summary of co-located QTLs and annotated functions of EST markers (SxM and OWB populations) and DArT markers (ICARDA population).** Investigated traits and annotated functions were generalised to the corresponding three different topics.

Transcription analysis of Bohnert *et al.* (2001), Seki *et al.* (2001) and Zhu *et al.* (2001) revealed that stress induced genes can be classified into two groups: genes expressing regulatory proteins and functional proteins. Regulatory proteins comprises transcription factors (DREB, ERE), protein kinases (MAPK, receptor protein kinases, ribosomal–protein kinases) and proteinases (phosphoesterases) involved in the regulation of signal transduction and gene expression. The group of functional proteins are membrane proteins and water channel proteins which maintain the water movement through the membrane. Furthermore also key enzymes for osmolyte biosyntheses, detoxification enzymes (glutathione *S*-transferase, hydrolase, catalase, superoxide dismutase and ascorbate peroxidase) and other proteins dedicated to macromolecule protection (LEA proteins, chaperons, mRNA binding) belong to this category (Agarwal *et al.* 2006). The annotated functions of detected markers in the second and third paper were classified into this scheme (Tab. 3) and selected proteins discussed in relation with seed longevity.

Molecular and genomic analyses display that several transcriptional regulatory systems are involved in the stress responsive induction of genes (Agarwal *et al.* 2006). Thereby the dehydration-responsive element-binding proteins (DREBs) and ethylene-responsive element (ERE) binding factors are two important subfamilies of the APETALA2/ethylene-responsive element-binding protein family (Sakuma *et al.* 2002) and play crucial roles in the regulation of abiotic- and biotic-stress responses, respectively (Sun *et al.* 2008). They belong to a large family of transcription factors which regulate gene expression in response to stress (Sakuma *et al.* 2002). Both are closely related to each other, but their target DNA binding sites are different which accompanies a functional diversity (Sun *et al.* 2008). The expression of different DREB homologs was mainly induced by cold, dehydration and high salt stresses which activated the transcription of stress-inducible genes (Shinozaki & Yamaguchi-Shinozaki 2000, Dubouzet *et al.* 2003). The accumulation of ROS activates ERE (Oracz *et al.* 2009) and enhances further the ethylene-dependent transcription which influences as endogenous plant hormone many aspects of plant development such as germination, senescence, abscission and fruit ripening (Ohme-Takagi & Shinshi 1995). Additionally ROS has putative roles in the influence on the initial seed quality and deterioration characteristics during seed storage.

Tab. 2: Classification of detected proteins according to the scheme of Agarwal *et al.* (2006).

Regulatory proteins	Functional proteins
APETALA2-like protein	Heat-shock protein
Dehydration responsive element binding protein (DREB)	Thaumatin-like protein TLP5
Ethylene-responsive element-binding factor 1 (EREBP-1)	Barperm1
Ethylene responsive element binding protein (ERE)	V-type ATPase
Putative NBS-LRR type resistance protein	VHA-A2 (Vacuolar proton ATPase A2]
Leucine rich repeat family protein	VHA-A3 (Vacuolar proton ATPase A3]
Putative serine/threonine-specific protein kinase	SUS5; UDP-glycosyltransferase/ sucrose synthase
Serine/threonine kinase-like protein	SUS6; UDP-glycosyltransferase/ sucrose synthase
NB-ARC domain containing protein	RNaseH
Putative light repressible receptor protein kinase	Putative permease 1
WD40-like beta propeller repeat family protein	Ribosomal protein P1
Putative HIRA	Enoyl-ACP reductase
	Beta-hordothionin precursor
	Alpha-2-purothionin precursor
	Putative asparate aminotransferase
	Stem rust resistance protein RPG1
	Jacalin-like lectin domain containing protein
	Poly(A)-binding protein

The initial germination is also related to plant disease resistance proteins which are often equipped with a series of leucine-rich repeats (LRRs), a nucleotide-binding site (NBS) and a putative amino-terminal signalling domain and can be finally termed as NBS-LRR proteins. The LRRs act as a kind of protein interaction platform and regulatory module of protein activation (Belkhadir *et al.* 2004). It is assumed that LBS-LRR proteins recognize the presence of pathogens directly or indirectly and activate numerous resistance genes (Meyers *et al.* 2003). It is also hypothesised that these proteins guard plant targets against pathogen effector proteins which enhance the susceptibility of the host plant while recognition is absent (Dangl & Jones 2001).

The reaction of functional proteins can be diverse and target a variety of tissues in seeds. As most regulatory proteins, a majority of the functional proteins can only act in hydrated tissue and influence seed longevity prior to the glassy state. The expression of the regulatory proteins is indicated by the appearance of RNAse H, involved in the initiation of DNA replication (Nowotny *et al.* 2005), ribosomal protein P1 required for translation (Szick *et al.* 1998) and poly-A-binding proteins responsible for post-transcriptional modification by providing binding sites for mRNA (Mangus *et al.* 2003). The synthesis of necessary amino acids is ensured by the function of asparate aminotransferase which plays, as key enzyme, important roles in regulation carbon and nitrogen metabolism (Zhou *et al.* 2009).

Under stress condition as for example salinity, drought, cold and anoxia the expression of V-type ATPase is up-regulated. This membrane protein belongs to a proton-pump which energizes endo-membranes. It enables active transport through membranes and has an important role in the regulation of cytoplasmic ion homeostasis (Dietz *et al.* 2001). The uptake of oxidised glutathione (GSSG) which we investigated in the third paper was also observed to be regulated by this enzyme (Tommasini *et al.* 1993) and is assumed to have an acidifying effect on the cytoplasm.

A group of anti-fungal, pathogenesis-related proteins are thaumatin-like proteins which are produced in response to pathogen attack. These proteins are able to alter fungal cell membrane integrity and lead to an inhibition of fungal growth, spore lysis, reduction in spore number and/or reduced viability of germinated spores (Velazhahan & Muthukrishnan 2003, Tobias *et al.* 2007). They were also reported during abiotic stress conditions and are assumed to be involved in plant development (Petre *et al.* 2011). Another class are the small, highly basic, cysteine rich thionins which can be broken into four major types based on the origin of plant species. Hordethionin and purothionin belong to the type-1 thionins that have been found in barley and wheat, respectively. High levels of hordothionin can lead to the permeabilisation of the fungal membrane and thus to its rupture (Carlson *et al.* 2006). Additionally to this characteristic the jacalin-related proteins have been suggested to be involved in resistance against insects (Chisholm *et al.* 2000). Belonging to a group of lectins they have generally been shown to confer defensive properties against bacteria and fungi (Peumans & Van Damme 1995).

The inclusion of initial germination data in the second and third paper might be the reason for the major appearance of processes which influence seed longevity before the material entered the glassy state. However, during desiccation metabolic activities cease (Bailly *et al.* 2008, Kranner *et al.* 2010b) and seeds are subjected to a different kind of stress. Due to the appearance of functions and QTLs associated with stress response and signalling there is a strong indication that reactive oxygen species (ROS) are involved (Mittler 2002, Apel & Hirt 2004, Mittler *et al.* 2004) and seeds need to be equipped with defence mechanism against this oxidative stress. In their study of long-term (22 years) stored maize inbred lines using a candidate gene approach, Revilla *et al.* (2009) could identify genes responsible for antioxidant enzymes as for example superoxide dismutase (SOD) and catalase (CAT). Further biochemical investigations in sunflower seed revealed a depletion of SOD, CAT and also glutathione reductase during ageing and a consequent seed viability loss (Kibinza *et al.* 2006, Lehner *et al.* 2008). In beech (*Fagus sylvatica* L.) the contents of α-tocopherol and ascorbic acid were positively correlated with germination capacity after different storage periods (Pukacka & Ratajczak 2007). In clover (*Trifolium repens* L. and *T. pratense* L.) peroxidase and CAT activity decreased while SOD increased after 40 years of storage (Cakmak *et al.* 2010).

In this context CAT was the first antioxidant enzyme discovered and plays an important role in plant response to changes in environmental conditions and biotic challenges. Principally, it catalyses the reaction of two H_2O_2 molecules to water and oxygen (Mhamdi *et al.* 2010). However, it influences additionally the cell redox state that is mainly controlled by the major low-molecule weight antioxidant glutathione. CAT knock-out mutants lead to perturbation of the redox-state and induce defence genes and oxidative stress-responsive transcripts associated with glutathione (Queval *et al.* 2007).

In the literature, the tripeptide thiol glutathione has minor importance in relation to seed ageing. Nevertheless, its relevance in the plant metabolism derives from its role as a signalling molecule, as a major reservoir of non-protein reduced sulphur in plants and as a cellular protectant (Noctor *et al.* 1998) which is realised by the antioxidative activity, the conjugation to electrophilic molecules by glutathione *S*-transferase (GST) and the detoxification of heavy metals (Foyer *et al.* 2001, Kranner & Colville 2010).

Glutathione (GSH), as shown in Fig. 12, is synthesised from glutamate, cysteine and glycine in two ATP-dependent reactions catalysed by γ-glutamylcysteine synthetase and glutathione synthetase (Foyer *et al.* 2001). To gain an overview how the different molecules react we performed an HPLC analyses to study cysteine, γ-glutamylcysteine, cysteinylglycine, oxidised and reduced glutathione in artificial aged and long-term stored barley seed.

According to Rauser *et al.* (1991) the major parts of GSH are located in the scutellum, where it protects directly the embryo against oxidative stress. Cysteine is predominated in the endosperm of maize and will be

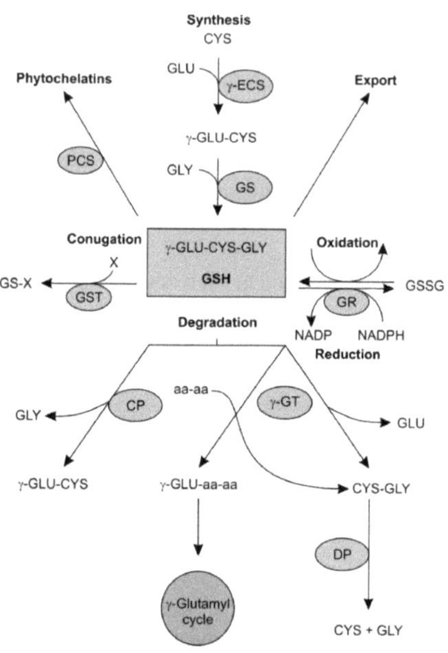

Fig. 6: Synthesis, use and degradation of glutathione in plant cells according Foyer *et al.* (2001). GSH, reduced glutathione; GSSG, oxidised glutathione; CYS, cysteine; GLU, glutamate; GLY, glycine; γ-GLU-CYS, γ-glutamylcysteine, CYS-GLY, cysteinylglycine; γ-ECS, γ-glutamylcysteine synthetase; GS, glutathione synthase; GR, glutathione reductase; γ-GT, γ-glutamyl transpeptidase DP, dipeptidase; GST, glutathione *S*-transferase; GS-X, conjugates; PCS, phytochelatins; aa-aa, dipeptides

absorbed as soon as metabolic activity increases during germination. A similar distribution is assumed in barley. However, due to the hard and dry barley kernel the whole grain was investigated and an increase of γ-glutamylcysteine and a depletion of cysteinylglycine and glutathione after ageing was found. The cysteine pool was mostly unaffected which indicates a passive contribution to the antioxidative activity of glutathione.

The total amount of glutathione decreases during storage which is caused by the water limitation that inactivates glutathione reductase and leads to the conjugation of glutathione molecules with storage proteins. The activity decline appears simultaneously with an increase of the GSSG/GSH ratio (Rhazi *et al.* 2003) and of the half-cell reduction potential $E_{GSSG/2GSH}$ which was also observed in the present barley study (third paper) after different ageing treatments. Especially an increase of $E_{GSSG/2GSH}$ is assumed to be part of a signalling cascade which triggers programmed cell death (PCD) and becomes lethal when $E_{GSSG/2GSH}$ exceeds -160 mV (Kranner *et al.* 2006). Interestingly the third paper shows that ageing treatments with increasing moisture content shift the lethal zone to more negative values which seems to be related to the acidification of the cytoplasm. This aspect needs to be considered in the Nernst equation when $E_{GSSG/2GSH}$ is applied as a universal stress marker (Kranner *et al.* 2006).

Under physiological conditions the ratio of oxidised to reduced glutathione is low (Foyer *et al.* 2001) but it rises during storage and depletes after seeds are imbibed and glutathione reductase can catalyse the reduction of GSSG (Kranner & Grill 1993). Seeds subjected to wet-dry cycles as in field conditions are able to restore frequently their antioxidative capacity and to maintain high viabilities (Long *et al.* 2011). Generally, for the generation of GSSG the molecule must be transferred to the cytosol (Foyer *et al.* 2001) which involves ATP-dependent transporters (Swanson *et al.* 1998) that has been detected in the genetic analyses and might differ between genotypes.

Similar to glutathione, the lipophilic antioxidants tocopherols and tocotrienols, commonly known as tocochromanols (Falk & Munné-Bosch 2010) scavenge various ROS but also lipid soluble by-products of oxidative stress and especially tocopherols undergo a cycling to restore their functions at the hydrated state (Sattler *et al.* 2004). They are relevant in preventing the propagation of peroxidation (Smirnoff 2010) and in the protection of membranes (Ratajczak & Pukacka 2005). They are proposed to have important roles in seed longevity and maintaining viability in *Arabidopsis* (Sattler *et al.* 2004), in herbaceous seepweed (*Suaeda maritima* L.) (Seal *et al.* 2010) and in beech seeds (Ratajczak & Pukacka 2005). Other studies have shown no correlation between tocopherols and viability loss as exemplified in soybean by Priestley *et al.* (1980). This result can be confirmed by the current study in barley. Neither the content of $\alpha, \beta, \gamma, \delta$-tocopherols nor the larger quantity of $\alpha, \beta, \gamma, \delta$-tocotrienols changed after long-term storage or artificial ageing. Barley seeds are endospermic and contain little oil as compared to the much oilier *Arabidopsis* seeds. Therefore, it will be interesting to test in future studies whether the responses of tocochromanols are related to oil content and/or seed morphology.

The major seed storage reserve showed a comparable pattern. The total amount of starch, oil and protein content did not reveal any change with ageing treatment. Therefore, a simple relationship can be excluded (Steadman *et al.* 1996). However, anecdotal accounts on the poor keeping quality of lettuce or peanut seeds have probably led to the idea that high oil content seeds are poorly storable (Walters *et al.* 2005a). Additional the shift in soluble carbohydrates during later stages of seed maturation and germination result in the hypotheses that soluble oligosaccharides play some role in seed longevity (Horbowicz & Obendorf 1994, Corbineau *et al.* 2000, Walters *et al.* 2005a). Studies of Sun *et al.* (1995), Bentsink *et al.* (2000) and Piotrowicz-Cieslak *et al.* (2010) did not support this assumption. Also Walters *et al.* (2005a) surveyed 118 species and could not find correlations between seed longevity and total sugar, oligosaccharides, sucrose content or soluble proteins. Only slightly significant correlations between seed lipid content and P50 could be assigned (Walters *et al.* 2005a, Nagel & Börner 2010) but were not confirmed in the current barley study.

In conclusion, biochemical reactions of seed ageing include a variety of processes which are triggered by different stresses and lead to the appearance of ROS including cascades of signalling and regulatory events at transcription level. However, in the glassy state antioxidants represent the main defence mechanism and deplete usually with seed ageing.

5.4 Application in genebanks and outlook

Since the beginning of the 20th century a lot of effort has been put into collecting unique genotypes from all over the world. Today, about 7.4 million accessions (FAO 2010) are safely stored in vaults but the challenge is now to keep them alive. One of the major pitfalls in seedbanks is the viability testing which is labour-, time- and cost-intensive. Nevertheless it is one of the most important seed properties that dictate the regeneration necessity. The present study could prove that similar handling of accessions during regeneration and storage will result in a wide range of viabilities after long-term storage caused by genetic or environmental effects. However, the differences in ageing rates result in a necessity of short monitoring frequencies which is expensive and depletes valuable seed samples in genebanks.

A variety of methods testing seed characteristics have already been developed and are summarised by Kranner (2010) but the application of especially non-invasive viability detection methods would be most beneficial for seedbanks. Unfortunately, the determination whether a seed is alive or dead has still similar limitations as pinpointing the location of an electron which can only be inferred after it has moved. Meaning the seed viability status is known as soon as the seed germinates (Walters 1998). At present infrared-thermography (Kranner *et al.* 2010a) and delayed luminescence (Yan *et al.* 2003, Costanzo *et al.* 2008) provide new opportunities for commercial development and application of non-invasive viability techniques, especially in seedbanks. Thereby, the infrared-thermography is based on a thermal profile of seed and image analyses which can detect imbibition and germination dependent biophysical and biochemical changes. The 'thermal fingerprint' varies with viability but also between species (Kranner *et al.* 2010a). Therefore adaptation to single seeds would be necessary. A totally different principle underlies delayed luminescence which measures single photons emitted by seeds. Dependent on the seed viability correlations were observed between artificially aged soybean (Costanzo *et al.* 2008) and barley (Yan *et al.* 2003) seed and the total number of photons emitted. Currently, both methods are at an experimental stage and further testing is necessary.

The present study carried out basic research using invasive techniques to unravel the genetic and biochemical secrets of seed ageing. The half-cell reduction potential $E_{GSSG/2GSH}$ and %GSSG revealed high correlations in relation with long-term stored and artificially aged seeds and in case of $E_{GSSG/2GSH}$ it has already been proposed as viability marker (Kranner *et al.* 2006). Unfortunately, the methodology has been still too expensive and time-consuming which excludes this method as quick viability measurement technique. Also the genome-wide association mapping study could be of interest for seed longevity estimations. In humans, a combination of significant SNP markers and a series of nested genetic risk models have been used to predict longevities. The analysis suggests that up to 150 SNPs should be sufficient to predict exceptional longevities with 77% accuracy in an independent set of centenarians and controls (Sebastiani *et al.* 2010). This would be also conceivable in barley seed where several loci have already been verified to be integrated in ageing processes.

Tab. 3: Invasive, semi-invasive and non-invasive diagnostic methods of seed viability detection according to Kranner (2010).

	Methods	Integrated processes, cell compartments and comments	Literature
Invasive	Germination test	Germination	ISTA (2008)
	Histochemical colourations (TTC assay, DAPI)	Viability, DNA	ISTA (2008) , Oracz *et al.* (2009)
	Spectrophotometry, HPLC, GC	Damages to lipids, proteins, DNA	Kranner *et al.* (2006)
	Electrophoresis	DNA-integrity (DNA-laddering); protein modification; functionality of defence and repair mechanism	ISTA (2008) Kranner *et al.* (2006)
Semi-invasive	Electric conductivity	Vigour test, applicable only in specific species; induction of anoxia, influence of metabolism	Hampton & TeKrony (1995) Demir *et al.* (2008)
	Micro-calorimetry	Metabolic heat; seed need to be equilibrated, therefore the first stage cannot be measured	Prat (1952), Hageseth & Cody (1993), Edelstein et al. (2001)
Non-invasive	Computer-assisted tomography in combination with image analyses	Seed morphology, size, colour, identification of empty seeds	Dell'Aquila (2007)
	Chlorophyll fluorescence of seed coat	Seed maturity	Jalink *et al.* (1999)
	Infrared spectroscopy	Protein, oil content, water content	Gergely & Andras (2003), van Loggerenberg & Pretorius (2005)
	Magnetic resonance, imaging	Seed morphology, dynamic illustration of water and oil content during imbibition	Terskikh *et al.* (2005a)
	Delayed luminescence	Germination	Costanzo *et al.* (2008), Yan *et al.* (2003)
	Nuclear magnetic resonance (NMR)	Metabolite-profiling (mobilisation of seed storage compounds)	Terskikh *et al.* (2005b)
	GC-MS	Vigour, volatile hydrocarbons and aldehydes	Tammela *et al.* (2003)
	Infrared thermography	Metabolic and non-metabolic heat (correlation with viability)	Kranner *et al.* (2010a)

However, up to now there are still some limitations in the presented methods that are related to artificial ageing of seeds at high moisture and high temperature levels. Therefore, another still ongoing project establishes an association study of wheat using long-term stored material from the Gatersleben seedbank and takes the advantage of slowly and dry aged material. A similar study will be set up in barley which enables the possibilities to compare loci with artificially aged material and to use the vast amount of information available for barley.

Further limitations of artificial ageing are assumed to be reduced by using a new approach with increased oxygen levels. Groot *et al.* (2011) developed methods that stimulate seed ageing by elevated partial pressure of oxygen (EPPO) at 18 MPa. The first trials using lettuce revealed similar damages of seedlings that have been shown in seeds after a few years of storage. Further examples of cabbage, soybean and pepper showed also effects within few weeks which allow using this method for gaining new information within a short period. The testing of the current barley mapping populations will show if loci are comparable under this condition.

6 Summary

With beginning of agriculture about 10,000 years ago, Neolithic farmers exerted a strong selection pressure on the wild progenitors of our today's crops and initialised a rise of cultivated plant species. Against the background of climate change, population growth, changes of agricultural systems and the green revolution, the genetic basis of these crops have become threatened during the last century. Single persons, institutes and conferences have dedicated their work to the valuable plant material and supported the establishment of genebanks and protected areas. Unfortunately the safekept material is again endangered due to an unknown viability status of most seed collections in genebanks.

The main objective of these studies were to investigate genetic and biochemical mechanisms underlying barley (*Hordeum vulgare* L.) seed deterioration with respect to genetic diversity at different storage treatments ranging from cold storage with low seed moisture content (smc) to heat treatment with high smc.

The first study was intended to provide viability information about selected cold and long-term stored crop species. Seeds of up to 50 genotypes of barley, wheat (*Triticum aestivum* L.), rye (*Secale cereale* L.), sorghum (*Sorghum bicolor* L.), flax (*Linum usitatissimum* L.) and oilseed rape (*Brassica napus* L.) were germinated after a minimum period of 26 years and revealed a spread of germination results which were smallest (60 to 100%) in case of sorghum and flax and widest (0 to 98%) in case of wheat and oilseed rape. Due to different initial germinations of the seed material and an unknown history concerning growth conditions and seed moisture content during storage, a strong environmental impact cannot be excluded. However, the absence of correlations between final and initial germinations as well as between final germination and seed moisture content point to a genetic component of germination separation and seed survival.

In order to clarify the genetic impact on seed deterioration quantitative genetic analyses using four mapping populations were applied. Caryopses, henceforth termed seeds, of three bi-parental barley mapping populations were artificially aged at 18 and 24% smc as well as 44 and 43°C, respectively, for 72 hours. Subsequent quantitative trait locus (QTL) analyses revealed 4 major loci on chromosomes 2H, 5H and 7H explaining a phenotypic variation up to 54%. Detected loci were confirmed by the fourth population that compromises a collection of independent barley accessions. These genotypes were multiplied in two field plots and artificially aged by low moisture contents (11% smc, 45°C, 16 days) and high moisture contents (18% smc, 43°C, 72 hours). Association analyses resulted in 105 marker-trait associations (MTAs) at 32 loci for germination (also termed vigour, abnormal seedlings excluded) and 98 MTAs at 28 loci for total germination (abnormal seedlings included). Four regions were comparable with seed longevity QTLs found in rice (*Oryza sativa* L.). Previously reported loci and published markers in close vicinity to the MTAs could be linked with agronomic traits and putative functions which were connected to plant characteristics as husk, plant height and to stress response as well as germination. The close connection between seed longevity loci and regulatory mechanism of abiotic and biotic stress response symbolises a dependency on environmental conditions during development to survive long-term storage.

The occurrences of genetic markers related to these stresses indicate a disruption of the cellular homeostasis that enhances the production of reactive oxygen species. The major antioxidant glutathione (GSH) is generally affected by oxidative stress and was analysed in 26 accessions which were maintained in parallel at cold (0°C) and ambient (20°C) storage up to 15 years. Further six of them (harvested in 2008) were artificially aged with low (13%) and high (18%) smc. In all storage treatments general depletion of the total glutathione pool and a shift towards more oxidising conditions, expressed by the oxidised glutathione (GSSG) and the half-cell reduction potential ($E_{GSSG/2GSH}$), were seen with ageing. The drier seeds of ambient and cold storage had a higher GSSG concentration than the wetter seeds exposed to artificial ageing, suggesting that enzymes were not active in the dry state to reduce GSSG to GSH. Artificial ageing at 18% smc was less destructive and enzymes could synthesize further GSH. Additionally the accumulation of γ-glutamylcysteine with vigour loss was significantly higher at ambient, cold storage and ageing at 13% smc indicating a partial degradation of GSH to the disulphide.

Comparing six barley genotypes over different storage and ageing treatments showed that genotypes maintained germination but with different concentrations of total GSH which is assumed to be specific to the genotype and may depend on the pre-harvest conditions. Therefore highly significant correlations over all treatments were not found for GSH but for percentage of GSSG of the total glutathione pool and $E_{GSSG/2GSH}$ which both represent good viability markers. Testing the same material on tocochromanols and storage compounds namely oil content, starch and proteins revealed that these metabolites were not affected by ageing treatments.

Current studies demonstrated that seed deterioration is dependent on a variety of factors which are associated with genotypic and environmental effects. Especially genetic analyses showed that cascades of signalling and stress response mechanism are activated which are assumed to be triggered by oxidative stress. This aspect became visible after analysis of glutathione and $E_{GSSG/2GSH}$ which decreased with ongoing long-term and artificial ageing treatment. Nevertheless a clear difference in glutathione behaviour was shown between these ageing treatments assuming a different kind of stress response. Therefore it is necessary to investigate long-term stored material because the results can provide new solutions for maintenance of unique genotypes in genebanks.

Zusammenfassung

Mit dem Beginn der Landwirtschaft vor etwa 10.000 Jahren begannen Neolithische Bauern den ersten Selektionsdruck auf unsere heutigen Kulturpflanzen auszuüben, womit sie den Grundstein der Pflanzendomestikation legten. Vor dem Hintergrund von Klimawandel, Bevölkerungswachstum, Veränderung landwirtschaftlicher Produktionssysteme und der grünen Revolution wurde die genetische Basis dieser Kulturpflanzen während des letzten Jahrhunderts stark eingeengt und gefährdet. Aus diesem Grund begannen Institute, Konferenzen und Personen ihre Arbeit den wertvollen pflanzengenetischen Ressourcen zu widmen und die Einrichtung von Genbanken und Schutzzonen zu unterstützen. Dennoch ist das sicher verwahrte Pflanzenmaterial aufgrund unbekannter Lebensfähigkeiten der meisten Genbankkollektionen in erneuter Gefahr.

Die Hauptziele der vorliegenden Studien bestehen darin, sowohl genetische als auch biochemische Mechanismen der Saatgutalterung von Kulturgerste (*Hordeum vulgare* L.) zu untersuchen, die das Verhalten genetisch diverser Akzessionen während unterschiedlicher Alterungssysteme, einschließlich Kühllagerung mit niedrigen Samenfeuchtegehalten (SFG) und Hitzebehandlung mit hohen SFG, berücksichtigen.

Zunächst erfolgte die Untersuchung von Samen ausgewählter kühl- und langzeitgelagerter Kulturpflanzenarten, um Informationen über deren Lebensfähigkeitsstatus zu erheben. Hierzu wurden bis zu 50 Gersten-, Weizen- (*Triticum aestivum* L.), Roggen- (*Secale cereale* L.), Sorghum- (*Sorghum bicolour* L.), Lein- (*Linum usitatissimum* L.) und Rapsgenotypen (*Brassica napus* L.) auf ihre Keimfähigkeit nach einer Lagerungszeit von mindestens 26 Jahren geprüft. Diese resultierten in weiten Keimfähigkeitsverteilungen, wobei Sorghum und Lein geringe (60-100%) und Weizen sowie Raps weite (0-98%) Spannbreiten umfassten. Aufgrund unterschiedlicher Anfangskeimfähigkeiten und der unbekannten Aufwuchs-, Ernte- und Lagerungshistorie des Untersuchungsmaterials kann ein starker Umwelteinfluss nicht ausgeschlossen werden. Jedoch deuten fehlende Korrelationen zwischen Anfangs- und Endkeimfähigkeiten sowie zwischen Endkeimfähigkeit und dem finalen SFG auf eine genetische Komponente der Keimungsunterschiede und demzufolge der Samenlanglebigkeit hin.

Um den genetischen Einfluss auf die Saatgutalterung darzustellen, wurden quantitative genetische Analysen unter Verwendung von vier Kartierungspopulationen durchgeführt. Gerstenkaryopsen, fortan als Samen bezeichnet, von drei bi-parentalen Kartierungs-populationen wurden künstlich bei 18 und 24% SFG beziehungsweise bei 44 und 43°C jeweils für 72 Stunden gealtert. Eine anschließende Quantitative Trait Locus (QTL) Analyse ergab insgesamt vier Genorte auf Chromosom 2H, 5H und 7H, die bis zu 54% der phänotypischen Variation erklärten. Die detektierten Genorte wurden weiterhin durch eine vierte Population bestätigt, die eine Kollektion von unabhängigen Gerstenakzessionen umfasste. Jene wurden in zwei gesonderten Vermehrungen angebaut und künstlich bei geringen Feuchtegehalten (11% SFG, 45°C, 16 Tage) und hohen Feuchtegehalten (18% SFG, 43°C, 72 Stunden) gealtert. Die Assoziationsanalyse ergab insgesamt 105 Marker-Trait-Assoziationen (MTAs) auf 32 Genorten für das Merkmal Keimfähigkeit (auch als Vigour bezeichnet, welches anomale Keimlinge ausschließt) und 98 MTAs auf 28 Genorten für das Merkmal Gesamt-Keimfähigkeit. Vier Genorte waren mit bekannten Positionen für

Samenlanglebigkeits-QTLs in Reis (*Oryza sativa* L.) vergleichbar. Zuvor erwähnte Genorte und veröffentlichte Marker nahe den MTAs konnten in Beziehung zu agronomischen Merkmalen mit möglichen Einflüssen auf Pflanzeneigenschaften, wie zum Beispiel Bespelzung, Pflanzenhöhe, Stressreaktion sowie Keimungsphysiologie gebracht werden. Die enge Beziehung zwischen Samenlanglebigkeit und genetisch basierten Regulierungsmechanismen von biotischen und abiotischen Stressreaktionen stellt eine enge Abhängigkeit von Umweltbedingungen während der Samenentwicklung und der Langlebigkeit dar.

Das Vorkommen dieser genetischen Marker in Verbindung mit Stress weist auf die Zerstörung der zellulären Homöostasis hin, welche die Produktion von reaktiven Sauerstoffspezies begünstigt. Das bedeutende Antioxidans Glutathion (GSH) ist generell durch oxidativen Stress beeinflusst und wurde infolgedessen in 26 gleichzeitig kühl ($0°C$) und ambient ($20°C$) gelagerten Akzessionen nach bis zu 15 Jahren untersucht. Weitere sechs dieser Akzessionen (Ernte 2008) wurden künstlich mit niedrigem (13%) und hohem (18%) SFG gealtert. Alle Alterungsbehandlungen zeichneten sich mit zunehmender Dauer durch eine signifikante Abnahme des Gesamtglutathiongehaltes und eine Verschiebung zu oxidativen Bedingungen aus, dargestellt durch oxidiertes Glutathion (GSSG) und das Reduktionspotential ($E_{GSSG/2GSH}$). Die trockeneren Samen der ambienten und kühlen Lagerung besaßen eine höhere GSSG-Konzentration als die feuchteren Samen der künstlichen Alterung, was auf eine fehlende Enzymaktivität im trockenen Zustand schließen lässt, welche GSSG anderenfalls zu GSH reduzierten. Die künstliche Alterung bei 18% war weniger destruktiv, was darauf schließen lässt, dass Enzyme weiteres GSH synthetisieren. Zusätzlich weist die Anhäufung von γ- Glutamylcystein während des GesamtKeimfähigkeitsverlustes bei ambienter und kühler Erhaltung sowie Lagerung bei 13% SFG auf eine teilweise Degradation von GSH zum Disulfid hin.

Der Vergleich von sechs Gerstenakzessionen über verschiedene Alterungsbehandlungen zeigt, dass die Genotypen bei gleichwertigen Keimfähigkeiten unterschiedliche Konzentrationen an Gesamtglutathion enthalten, welches auf ein Genotyp spezifisches Verhalten schließt, eventuell beeinflusst durch Vorerntebedingungen. Aus diesem Grund ergab der Gesamtglutathionsgehalt keine signifikante Korrelation über alle Alterungsbehandlungen, hingegen korrelierten der prozentuale Anteil von GSSG am Gesamtglutathion und $E_{GSSG/2GSH}$, womit beide Parameter gute Lebensfähigkeitsmarker darstellen. Die Analyse desselben Materials auf Tocochromanole und Reservestoffe, insbesondere Fette, Stärke und Eiweiße ergab keine Beeinträchtigung dieser Metaboliten durch die Alterungsbehandlungen.

Die gegenwärtigen Studien zeigen, dass die Samenalterung von einer Reihe Faktoren abhängig ist, die mit genotypischen und Umwelteffekten verbunden sind. Insbesondere die genetische Analyse gibt Hinweise auf eine Kaskade an Signal- und Stressreaktionsmechanismen, die unter anderem durch oxidativen Stress ausgelöst werden können. Dieser Aspekt wurde durch die Analyse von Glutathion und $E_{GSSG/2GSH}$ deutlich, da sich mit zunehmender Alterungsdauer Glutathion signifikant verringerte. Ein deutlicher Unterschied wurde zwischen den Alterungsbehandlungen nachgewiesen, die unterschiedliche Stressreaktionen vermuten lassen. Aus diesem Grund ist es notwendig Untersuchungen an langzeitgelagertem Material durchzuführen, um Genbanken neue Lösungswege für die Erhaltung von einzigartigen Genotypen präsentieren zu können.

7   References of Prolog and Epilog

Agarwal P., P. Agarwal, M. Reddy & S. Sopory. (2006) Role of DREB transcription factors in abiotic and biotic stress tolerance in plants. Plant Cell Reports 25, 1263-1274.

Akbari M., P. Wenzl, V. Caig, J. Carling, L. Xia, S.Y. Yang, G. Uszynski, V. Mohler, A. Lehmensiek, H. Kuchel, M.J. Hayden, N. Howes, P. Sharp, P. Vaughan, B. Rathmell, E. Huttner & A. Kilian. (2006) Diversity arrays technology (DArT) for high-throughput profiling of the hexaploid wheat genome. Theoretical and Applied Genetics 113, 1409-1420.

Alonso-Blanco C., M.G.M. Aarts, L. Bentsink, J.J.B. Keurentjes, M. Reymond, D. Vreugdenhil & M. Koornneef. (2009) What has natural variation taught us about plant development, physiology, and adaptation? The Plant Cell Online 21, 1877-1896.

Alpert P. (2005) The limits and frontiers of desiccation-tolerant life. Integrative and Comparative Biology 45, 685-695.

Alpert P. (2006) Constraints of tolerance: Why are desiccation-tolerant organisms so small or rare? Journal of Experimental Biology 209, 1575-1584.

Angell C.A. (1991) Relaxation in liquids, polymers and plastic crystals - strong/ fragile patterns and problems. Journal of Non-Crystalline Solids 131-133, 13-31.

Apel K. & H. Hirt. (2004) Reactive oxygen species: Metabolism, oxidative stress, and signal transduction. Annual Review of Plant Biology 55, 373-399.

Aufhammer G. & U. Simon. (1957) Die Samen landwirtschaftlicher Kulturpflanzen im Grundstein des ehemaligen Nürnberger Stadttheaters und ihre Keimfähigkeit. Zeitschrift für Acker- und Pflanzenbau 103, 455-472.

Avrami M. (1941) Granulation, phase change, and microstructure kinetics of phase change III. The Journal of Chemical Physics 9, 177-184.

Baik B.-K. & S.E. Ullrich. (2008) Barley for food: Characteristics, improvement, and renewed interest. Journal of Cereal Science 48, 233-242.

Bailly C., H. El-Maarouf-Bouteau & F. Corbineau. (2008) From intracellular signaling networks to cell death: The dual role of reactive oxygen species in seed physiology. Comptes Rendus Biologies 331, 806-814.

Barzali M., U. Lohwasser, M. Niedzielski & A. Börner. (2005) Effects of different temperatures and atmospheres on seed and seedling traits in a long-term storage experiment on rye (*Secale cereale* L.). Seed Science and Technology 33, 713-721.

Baskin J.M. & C.C. Baskin. (2004) A classification system for seed dormancy. Seed Science Research 14, 1-16.

Bassham D.C. (2007) Plant autophagy - more than a starvation response. Current Opinion in Plant Biology 10, 587-593.

Belkhadir Y., R. Subramaniam & J.L. Dangl. (2004) Plant disease resistance protein signaling: NBS-LRR proteins and their partners. Current Opinion in Plant Biology 7, 391-399.

Bentsink L., C. Alonso-Blanco, D. Vreugdenhil, K. Tesnier, S.P.C. Groot & M. Koornneef. (2000) Genetic analysis of seed-soluble oligosaccharides in relation to seed storability of *Arabidopsis*. Plant Physiology 124, 1595-1604.

Bergen P.F.v., H.A. Bland, M.C. Horton & R.P. Evershed. (1997) Chemical and morphological changes in archaeological seeds and fruits during preservation by desiccation. Geochimica Et Cosmochimica Acta 61, 1919-1930.

Berjak P. & N.W. Pammenter. (2008) From *Avicennia* to *Zizania*: Seed recalcitrance in perspective. Annals of Botany 101, 213-228.

Bewley J.D. (1997) Seed germination and dormancy. Plant Cell 9, 1055-1066.

Bhullar N.K., K. Street, M. Mackay, N. Yahiaoui & B. Keller. (2009) Unlocking wheat genetic resources for the molecular identification of previously undescribed functional alleles at the *Pm3* resistance locus. Proceedings of the National Academy of Sciences 106, 9519-9524.

Bioversity. (2011) History of Bioversity. http://www.bioversityinternational.org/about_us/ history_of_bioversity.html, Bioversity International, Rome, visited at 04th April 2011.

Bohnert H.J., P. Ayoubi, C. Borchert, R.A. Bressan, R.L. Burnap, J.C. Cushman, M.A. Cushman, M. Deyholos, R. Fischer, D.W. Galbraith, P.M. Hasegawa, M. Jenks, S. Kawasaki, H. Koiwa, S. Kore-Eda, B.-H. Lee, C.B. Michalowski, E. Misawa, M. Nomura, N. Ozturk, B. Postier, R. Prade, C.-P. Song, Y. Tanaka, H. Wang & J.-K. Zhu. (2001) A genomics approach towards salt stress tolerance. Plant Physiology and Biochemistry 39, 295-311.

Börner A. (2006) Preservation of plant genetic resources in the biotechnology era. Biotechnology Journal 1, 1393-1404.

Boubriak I., H. Kargiolaki, L. Lyne & D.J. Osborne. (1997) The requirement for DNA repair in desiccation tolerance of germinating embryos. Seed Science Research 7, 97-106.

Bray C.M. & C.E. West. (2005) DNA repair mechanisms in plants: Crucial sensors and effectors for the maintenance of genome integrity. New Phytologist 168, 511-528.

Buck-Sorlin G.H. (2002) The search for QTL in barley (*Hordeum vulgare* L.) using a new mapping population. Cellular & Molecular Biology Letters 7, 523-535.

Buitink J. & O. Leprince. (2004) Glass formation in plant anhydrobiotes: Survival in the dry state. Cryobiology 48, 215-228.

Buitink J., C. Walters, F.A. Hoekstra & J. Crane. (1998) Storage behavior of *Typha latifolia* pollen at low water contents: Interpretation on the basis of water activity and glass concepts. Physiologia Plantarum 103, 145-153.

Buitink J., O. Leprince, M.A. Hemminga & F.A. Hoekstra. (2000) Molecular mobility in the cytoplasm: An approach to describe and predict lifespan of dry germplasm. Proceedings of the National Academy of Sciences of the United States of America 97, 2385-2390.

Cakmak T., O. Atici & G. Agar. (2010) The natural aging-related biochemical changes in the seeds of two legume varieties stored for 40 years. Acta Agriculturae Scandinavica Section B-Soil and Plant Science 60, 353-360.

Cantino P.D., J. Doyle, S. Graham, W.S. Judd, R.G. Olmstead, D.E. Soltis, P.S. Soltis & M.J. Donoghue. (2007) Towards a phylogenetic nomenclature of *Tracheophyta*. Taxon 56, 822-846.

Carlson A., R. Skadsen & H. Kaeppler. (2006) Barley hordothionin accumulates in transgenic oat seeds and purified protein retains anti-fungal properties *in vitro*. In Vitro Cellular & Developmental Biology - Plant 42, 318-323.

CBD (2002) Global strategy for plant conservation Montreal, Canada, The Secretariat of the Convention on Biological Diversity.

CBD. (2010) International year of biodiversity 2010. The Secretariat of the Convention on Biological Diversity, http://www.cbd.int/2010/welcome/, Montreal, Canada, visited on 20th April 2011.

Chisholm S.T., S.K. Mahajan, S.A. Whitham, M.L. Yamamoto & J.C. Carrington. (2000) Cloning of the *Arabidopsis RTM1* gene, which controls restriction of long-distance movement of tobacco etch virus. Proceedings of the National Academy of Sciences 97, 489-494.

Chwedorzewska K.J., P.T. Bednarek, R. Lewandowska, P. Krajewski & J. Puchalski. (2006) Studies on genetic changes in rye samples (*Secale cereale* L.) maintained in a seed bank. Cellular & Molecular Biology Letters 11, 338-347.

Clerkx E.J.M., M.E. El-Lithy, E. Vierling, G.J. Ruys, H. Blankestijin-De Vries, S.P.C. Groot, D. Vreugdenhil & M. Koornneef. (2004) Analysis of natural allelic variation of *Arabidopsis* seed germination and seed longevity traits between the accessions Landsberg erecta and Shakdara, using a new recombinant inbred line population. Plant Physiology 135, 432-443.

Cockram J., H. Hones & D.M. O'Sullivan. (2011) Genetic variation at flowering time loci in wild and cultivated barley. Plant Genetic Resources FirstView, 1-4.

Colville L. & I. Kranner. (2010) Desiccation tolerant plants as model systems to study redox regulation of protein thiols. Plant Growth Regulation 62, 241-255.

Corbineau F., M.A. Picard, J.A. Fougereux, F. Ladonne & D. Come. (2000) Effects of dehydration conditions on desiccation tolerance of developing pea seeds as related to oligosaccharide content and cell membrane properties. Seed Science Research 10, 329-339.

Corvalan C., S. Hales & A. McMichael (2005) Ecosystems and human well-being: Biodiversity synthesis Millennium Ecosystem Assessment. Washington, DC, USA, World Resources Institute.

Costa J.M., A. Corey, P.M. Hayes, C. Jobet, A. Kleinhofs, A. Kopisch-Obusch, S.F. Kramer, D. Kudrna, M. Li, O. Riera-Lizarazu, K. Sato, P. Szucs, T. Toojinda, M.I. Vales & R.I. Wolfe. (2001) Molecular mapping of the Oregon Wolfe Barleys: A phenotypically polymorphic doubled-haploid population. Theoretical and Applied Genetics 103, 415-424.

Costanzo E., M. Gulino, L. Lanzano, F. Musumeci, A. Scordino, S. Tudisco & L. Sui. (2008) Single seed viability checked by delayed luminescence. European Biophysics Journal with Biophysics Letters 37, 235-238.

Crepet W.L. & K.J. Niklas. (2009) Darwin's second 'abominable mystery': Why are there so many angiosperm species? American Journal of Botany 96, 366-381.

Crowe J.H., F.A. Hoekstra & L.M. Crowe. (1992) Anhydrobiosis. Annual Review of Physiology 54, 579-599.

Damania A.B. (2008) History, achievements, and current status of genetic resources conservation. Agronomy Journal 100, S27-S39.

Dangl J.L. & J.D.G. Jones. (2001) Plant pathogens and integrated defence responses to infection. Nature 411, 826-833.

Davies M.J., S.L. Fu, H.J. Wang & R.T. Dean. (1999) Stable markers of oxidant damage to proteins and their application in the study of human disease. Free Radical Biology and Medicine 27, 1151-1163.

Daws M.I., J. Davies, E. Vaes, R. van Gelder & H.W. Pritchard. (2007) Two-hundred-year seed survival of *Leucospermum* and two other woody species from the Cape Floristic region, South Africa. Seed Science Research 17, 73-79.

De Gara L., V. Locato, S. Dipierro & M.C. de Pinto. (2010) Redox homeostasis in plants. The challenge of living with endogenous oxygen production. Respiratory Physiology & Neurobiology 173, S13-S19.

De Tullio M.C. & O. Arrigoni. (2003) The ascorbic acid system in seeds: To protect and to serve. Seed Science Research 13, 249-260.

Debeaujon I., K.M. Leon-Kloosterziel & M. Koornneef. (2000) Influence of the testa on seed dormancy, germination, and longevity in *Arabidopsis*. Plant Physiology 122, 403-413.

Dell'Aquila A. (2007) Towards new computer imaging techniques applied to seed quality testing and sorting. Seed Science and Technology 35, 519-538.

Delouche J.C. & C.C. Baskin. (1973) Accelerated aging techniques for predicting the relative storability of seed lots. Seed Science and Technology 1, 427-452.

Demir I., K. Mavi, B.B. Kenanoglu & S. Matthews. (2008) Prediction of germination and vigour in naturally aged commercially available seed lots of cabbage (*Brassica oleracea* var. *capitata*) using the bulk conductivity method. Seed Science and Technology 36, 509-523.

Dickie J.B., R.H. Ellis, H.L. Kraak, K. Ryder & P.B. Tompsett. (1990) Temperature and seed storage longevity. Annals of Botany 65, 197-204.

Dietz K.J., N. Tavakoli, C. Kluge, T. Mimura, S.S. Sharma, G.C. Harris, A.N. Chardonnens & D. Golldack. (2001) Significance of the V-type ATPase for the adaptation to stressful growth conditions and its regulation on the molecular and biochemical level. Journal of Experimental Botany 52, 1969-1980.

Dubouzet J.G., Y. Sakuma, Y. Ito, M. Kasuga, E.G. Dubouzet, S. Miura, M. Seki, K. Shinozaki & K. Yamaguchi-Shinozaki. (2003) *OsDREB* genes in rice, *Oryza sativa* L., encode transcription activators that function in drought-, high-salt- and cold-responsive gene expression. Plant Journal 33, 751-763.

Edelstein M., K.J. Bradford & D.W. Burger. (2001) Metabolic heat and CO_2 production rates during germination of melon (*Cucumis melo* L.) seeds measured by microcalorimetry. Seed Science Research 11, 265-272.

El-Maarouf-Bouteau H. & C. Bailly. (2008) Oxidative signaling in seed germination and dormancy. Plant Signaling & Behavior 3, 175-182.

El-Maarouf-Bouteau H., C. Job, D. Job, F. Corbineau & C. Bailly. (2007) ROS production in seed dormancy alleviation. Plant Signaling & Behavior 2, 362-364.

Ellis R.H. & E.H. Roberts. (1980a) The influence of temperature and moisture on seed viability period in barley (*Hordeum distichum* L.). Annals of Botany 45, 31-37.

Ellis R.H. & E.H. Roberts. (1980b) Improved equations for the prediction of seed longevity. Annals of Botany 45, 13-30.

Ertugay M.F. & M. Certel. (2000) Moisture sorption isotherms of cereals at different temperatures. Nahrung-Food 44, 107-109.

Evershed R.P., H.A. Bland, P.F.v. Bergen, J.F. Carter, M.C. Horton & P.A. RowleyConwy. (1997) Volatile compounds in archaeological plant remains and the Maillard reaction during decay of organic matter. Science 278, 432-433.

Falk J. & S. Munné-Bosch. (2010) Tocochromanol functions in plants: antioxidation and beyond. Journal of Experimental Botany 61, 1549-1566.

FAO (2010) The second report on the state of the world's plant genetic resources for food and agriculture. Rome.

FAO. (2011) FAOSTAT: Crops statistic 2009. Food and Agriculture Organization of the United Nations, http://faostat.fao.org/site/567/DesktopDefault.aspx?PageID=567# ancor, Rome, visited at 06th April 2011.

FAO/IPGRI (1994) Genebank Standards., Food and Agriculture Organization of the United Nations, Rome, International Plant Genetic Resources Institute, Rome.

Fath A., P.C. Bethke & R.L. Jones. (2001) Enzymes that scavenge reactive oxygen species are down-regulated prior to gibberellic acid-induced programmed cell death in barley aleurone. Plant Physiology 126, 156-166.

Finch-Savage W.E. & G. Leubner-Metzger. (2006) Seed dormancy and the control of germination. New Phytologist 171, 501-523.

Finch-Savage W.E., C.S.C. Cadman, P.E. Toorop, J.R. Lynn & H.W.M. Hilhorst. (2007) Seed dormancy release in *Arabidopsis* Cvi by dry after-ripening, low temperature, nitrate and light shows common quantitative patterns of gene expression directed by environmentally specific sensing. The Plant Journal 51, 60-78.

Finkel T. & N.J. Holbrook. (2000) Oxidants, oxidative stress and the biology of ageing. Nature 408, 239-247.

Finkelstein R., W. Reeves, T. Ariizumi & C. Steber. (2008) Molecular aspects of seed dormancy. Annual Review of Plant Biology 59, 387-415.

Finkelstein R.R., S.S.L. Gampala & C.D. Rock. (2002) Abscisic acid signaling in seeds and seedlings. The Plant Cell 14, S15-S45.

Forbis T.A., S.K. Floyd & A.d. Queiroz. (2002) The evolution of embryo size in angiosperms and other seed plants: Implications for the evolution of seed dormancy. Evolution 56, 2112-2125.

Foyer C.H. & G. Noctor. (2011) Ascorbate and glutathione: The heart of the redox hub. Plant Physiology 155, 2-18.

Foyer C.H., F.L. Theodoulou & S. Delrot. (2001) The functions of inter- and intracellular glutathione transport systems in plants. Trends in Plant Science 6, 486-492.

França M.B., A.D. Panek & E.C.A. Eleutherio. (2007) Oxidative stress and its effects during dehydration. Comparative Biochemistry and Physiology - Part A: Molecular & Integrative Physiology 146, 621-631.

Franks F., R.H.M. Hatley & S. Mathias. (1991) Materials science and the production of shelf-stable biologicals. BioPharm 4, 38-42.

Friedman W.E. (1995) Organismal duplication, inclusive fitness theory, and altruism: Understanding the evolution of endosperm and the angiosperm reproductive syndrome. Proceedings of the National Academy of Sciences 92, 3913-3917.

Friedman W.E. (2009) The meaning of Darwin's 'abominable mystery'. American Journal of Botany 96, 5-21.

Friedman W.E. & K.C. Ryerson. (2009) Reconstructing the ancestral female gametophyte of angiosperms: Insights from *Amborella* and other ancient lineages of flowering plants. American Journal of Botany 96, 129-143.

Gaut B.S. (2002) Evolutionary dynamics of grass genomes. New Phytologist 154, 15-28.

Gergely S. & A. András. (2003) Changes in moisture content during wheat maturation - what is measured by near infrared spectroscopy? Journal of Near Infrared Spectroscopy 11, 17-26.

Grando S. & H. Gormez Macpherson (2005) Food barley: Importance, uses and local knowledge Proceedings of the International Workshop on Food Barley Improvement, 14th-17th January 2002. Hammamet, Tunisia.

Greenberg J.T. (1996) Programmed cell death: A way of life for plants. Proceedings of the National Academy of Sciences 93, 12094-12097.

Groos C., G. Gay, M.R. Perretant, L. Gervais, M. Bernard, F. Dedryver & G. Charmet. (2002) Study of the relationship between pre-harvest sprouting and grain color by quantitative trait loci analysis in a white×red grain bread-wheat cross. Theoretical and Applied Genetics 104, 39-47.

Groot S.P.C., A. Abbasi Surki & J. Kodde. (2011) Elevated partial pressure of oxygen - a novel method to study seed ageing and longevity. Informativo Abrates 21, 104.

Haferkamp M.E., L. Smith & R.A. Nilan. (1953) Studies on aged seeds I. Relation of age of seed to germination and longevity. Agronomy Journal 45, 434-437.

Hageseth G.T. & A.L. Cody. (1993) Energy-level model for isothermal seed germination. Journal of Experimental Botany 44, 119-125.

Hall D., C. Tegstrom & P.K. Ingvarsson. (2010) Using association mapping to dissect the genetic basis of complex traits in plants. Briefings in Functional Genomics 9, 157-165.

Halliwell B. (2006) Reactive species and antioxidants. Redox biology is a fundamental theme of aerobic life. Plant Physiology 141, 312-322.

Hampton J.G. & D.M. TeKrony (1995) Handbook of vigour test methods. (3 edition) Zürich, International Seed Testing Association.

Harlan J.R. (1975) Crops and man. USA, Madison, Wis.: American Society of Agronomy.

Harrington J.F. (1963) Practical instructions and advice on seed storage. pp 989-994 Proceedings of the International Seed Testing Association

Hawkes J.G. (1983) The diversity of crop plants. Cambridge, MA, Harvard University Press.

Hay F.R., A. Mead, K. Manger & F.J. Wilson. (2003) One-step analysis of seed storage data and the longevity of *Arabidopsis thaliana* seeds. Journal of Experimental Botany 54, 993-1011.

Hayes P. & P. Szucs. (2006) Disequilibrium and association in barley: Thinking outside the glass. Proceedings of the National Academy of Sciences of the United States of America 103, 18385-18386.

Hengartner M.O. (2000) The biochemistry of apoptosis. Nature 407, 770-776.

Hilhorst H.W.M. (1995) A critical update on seed dormancy I. Primary dormancy. Seed Science Research 5, 61-73.

Hoekstra F.A. (2005) Differential longevities in desiccated anhydrobiotic plant systems. Integrative and Comparative Biology 45, 725-733.

Holland J.B. (2007) Genetic architecture of complex traits in plants. Current Opinion in Plant Biology 10, 156-161.

Hopkin M. (2007) Norway unveils design of 'doomsday' seed bank. Nature 445, 693-693.

Horbowicz M. & R.L. Obendorf. (1994) Seed desiccation tolerance and storability: Dependence on flatulence-producing oligosaccharides and cyclitols - review and survey. Seed Science Research 4, 385-405.

Hyten D., Q. Song, I.-Y. Choi, M.-S. Yoon, J. Specht, L. Matukumalli, R. Nelson, R. Shoemaker, N. Young & P. Cregan. (2008) High-throughput genotyping with the GoldenGate assay in the complex genome of soybean. Theoretical and Applied Genetics 116, 945-952.

Illing N., K.J. Denby, H. Collett, A. Shen & J.M. Farrant. (2005) The signature of seeds in resurrection plants: A molecular and physiological comparison of desiccation tolerance in seeds and vegetative tissues. Integrative and Comparative Biology 45, 771-787.

ISTA (2008) International Rules for Seed Testing. Bassersdorf, Switzerland, International Seed Testing Association.

Iturriaga G., R. Suárez & B. Nova-Franco. (2009) Trehalose metabolism: From osmoprotection to signaling. International Journal of Molecular Sciences 10, 3793-3810.

Jalink H., R. van der Schoor, Y.E. Birnbaum & R.J. Bino. (1999) Seed chlorophyll content as an indicator for seed maturity and seed quality. Acta Horticulturae (ISHS) 540, 219-228.

Jones A. (2000) Does the plant mitochondrion integrate cellular stress and regulate programmed cell death? Trends in Plant Science 5, 225-230.

Kets E.P.W., P.J. Ijpelaar, F.A. Hoekstra & H. Vromans. (2004) Citrate increases glass transition temperature of vitrified sucrose preparations. Cryobiology 48, 46-54.

KEW. (2008) Seed Information Database (SID). Version 7.1. Royal Botanical Gardens KEW, http://data.kew.org/sid/ (May 2008), Wakehurst Place, UK, visited at the 21st April 2011.

Khan M.M., G.A.F. Hendry, N.M. Atherton & C.W. Vertucci-Walters. (1996) Free radical accumulation and lipid peroxidation in testas of rapidly aged soybean seeds: A light-promoted process. Seed Science Research 6, 101-107.

Kibinza S., D. Vinel, D. Come, C. Bailly & F. Corbineau. (2006) Sunflower seed deterioration as related to moisture content during ageing, energy metabolism and active oxygen species scavenging. Physiologia Plantarum 128, 496-506.

Kilian B., W. Martin & F. Salamini. (2010) Genetic diversity, evolution and domestication of wheat and barley in the Fertile Crescent. pp 137-166 In: M. Glaubrecht (Ed) Evolution in Action, Springer Berlin Heidelberg.

Kislev M.E., D. Nadel & I. Carmi. (1992) Epipalaeolithic (19,000 BP) cereal and fruit diet at Ohalo II, Sea of Galilee, Israel. Review of Palaeobotany and Palynology 73, 161-166.

Kivilaan A. & R.S. Bandurski. (1981) The 100-year period for Beal seed viability experiment. American Journal of Botany 68, 1290-1292.

Kleinhofs A., A. Kilian, M.A. Saghai Maroof, R.M. Biyashev, P. Hayes, F.Q. Chen, N. Lapitan, A. Fenwick, T.K. Blake, V. Kanazin, E. Ananiev, L. Dahleen, D. Kudrna, J. Bollinger, S.J. Knapp, B. Liu, M. Sorrells, M. Heun, J.D. Franckowiak, D. Hoffman, R. Skadsen & B.J. Steffenson. (1993) A molecular, isozyme and morphological map of the barley (*Hordeum vulgare*) genome. Theoretical and Applied Genetics 86, 705-712.

Kochanek J., K.J. Steadman, R.J. Probert & S.W. Adkins. (2009) Variation in seed longevity among different populations, species and genera found in collections from wild Australian plants. Australian Journal of Botany 57, 123-131.

Komatsuda T., M. Pourkheirandish, C.F. He, P. Azhaguvel, H. Kanamori, D. Perovic, N. Stein, A. Graner, T. Wicker, A. Tagiri, U. Lundqvist, T. Fujimura, M. Matsuoka, T. Matsumoto & M. Yano. (2007) Six-rowed barley originated from a mutation in a homeodomain-leucine zipper I-class homeobox gene. Proceedings of the National Academy of Sciences of the United States of America 104, 1424-1429.

Koornneef M., L. Bentsink & H. Hilhorst. (2002) Seed dormancy and germination. Current Opinion in Plant Biology 5, 33-36.

Kranner H. & D. Grill. (1993) Content of low-molecular-weight thiols during the imbibition of pea seeds. Physiologia Plantarum 88, 557-562.

Kranner I. (2010) Erhaltung pflanzengenetischer Ressourcen in Genbanken – Physiologische und biochemische Grundlagen. Berichte der Gesellschaft für Pflanzenbauwissenschaften Band 5, 1-6.

Kranner I. & D. Grill. (1996) Significance of thiol-disulfide exchange in resting stages of plant development. Botanica Acta 109, 8-14.

Kranner I. & L. Colville. (2010) Metals and seeds: Biochemical and molecular implications and their significance for seed germination. Environmental and Experimental Botany 72, 93-105.

Kranner I., S. Birtic, K.M. Anderson & H.W. Pritchard. (2006) Glutathione half-cell reduction potential: A universal stress marker and modulator of programmed cell death? Free Radical Biology and Medicine 40, 2155-2165.

Kranner I., G. Kastberger, M. Hartbauer & H.W. Pritchard. (2010a) Noninvasive diagnosis of seed viability using infrared thermography. Proceedings of the National Academy of Sciences of the United States of America 107, 3912-3917.

Kranner I., F.V. Minibayeva, R.P. Beckett & C.E. Seal. (2010b) What is stress? Concepts, definitions and applications in seed science. New Phytologist 188, 655-673.

Kranner I., T. Roach, R.P. Beckett, C. Whitaker & F.V. Minibayeva. (2010c) Extracellular production of reactive oxygen species during seed germination and early seedling growth in *Pisum sativum*. Journal of Plant Physiology 167, 805-811.

Kranner I., H. Chen, H. Pritchard, S. Pearce & S. Birtić. (2011) Inter-nucleosomal DNA fragmentation and loss of RNA integrity during seed ageing. Plant Growth Regulation 63, 63-72.

Kranner I., W.J. Cram, M. Zorn, S. Wornik, I. Yoshimura, E. Stabentheiner & H.W. Pfeifhofer. (2005) Antioxidants and photoprotection in a lichen as compared with its isolated symbiotic partners. Proceedings of the National Academy of Sciences of the United States of America 102, 3141-3146.

Kucera B., M.A. Cohn & G. Leubner-Metzger. (2005) Plant hormone interactions during seed dormancy release and germination. Seed Science Research 15, 281-307.

Lam E., N. Kato & M. Lawton. (2001) Programmed cell death, mitochondria and the plant hypersensitive response. Nature 411, 848-853.

Lee G.J., N. Pokala & E. Vierling. (1995) Structure and *in vitro* molecular chaperone activity of cytosolic small heat shock proteins from pea. Journal of Biological Chemistry 270, 10432-10438.

Lehner A., N. Mamadou, P. Poels, D. Come, C. Bailly & F. Corbineau. (2008) Changes in soluble carbohydrates, lipid peroxidation and antioxidant enzyme activities in the embryo during ageing in wheat grains. Journal of Cereal Science 47, 555-565.

Leino M.W. & J. Edqvist. (2010) Germination of 151-year old *Acacia* spp. seeds. Genetic Resources and Crop Evolution 57, 741-746.

Leubner-Metzger G. (2005) β-1,3-Glucanase gene expression in low-hydrated seeds as a mechanism for dormancy release during tobacco after-ripening. The Plant Journal 41, 133-145.

Lev-Yadun S., A. Gopher & S. Abbo. (2000) The cradle of agriculture. Science 288, 1602-1603.

Levine B. & D.J. Klionsky. (2004) Development by self-digestion: Molecular mechanisms and biological functions of autophagy. Developmental Cell 6, 463-477.

Li D.Z. & H.W. Pritchard. (2009) The science and economics of *ex situ* plant conservation. Trends in Plant Science 14, 614-621.

Linkies A., K. Graeber, C. Knight & G. Leubner-Metzger. (2010) The evolution of seeds. New Phytologist 186, 817-831.

Loggerenberg D.v. & A. Pretorius. (2005) Determining oil and protein content of striped and black sunflower seeds using NIR technology. Communications in Soil Science and Plant Analysis 36, 729 - 741.

Long R., I. Kranner, F. Panetta, S. Birtic, S. Adkins & K. Steadman. (2011) Wet-dry cycling extends seed persistence by re-instating antioxidant capacity. Plant and Soil 338, 511-519.

Mackay I. & W. Powell. (2007) Methods for linkage disequilibrium mapping in crops. Trends in Plant Science 12, 57-63.

Mangus D., M. Evans & A. Jacobson. (2003) Poly(A)-binding proteins: multifunctional scaffolds for the post-transcriptional control of gene expression. Genome Biology 4, 223.

Mattson M.P. & S.L. Chan. (2003) Calcium orchestrates apoptosis. Nature Cell Biology 5, 1041-1043.

Meinhard M. & E. Grill. (2001) Hydrogen peroxide is a regulator of *ABI1*, a protein phosphatase 2C from *Arabidopsis*. FEBS Letters 508, 443-446.

Meinhard M., P. Rodriguez & E. Grill. (2002) The sensitivity of ABI2 to hydrogen peroxide links the abscisic acid-response regulator to redox signalling. Planta 214, 775-782.

Meyers B.C., A. Kozik, A. Griego, H. Kuang & R.W. Michelmore. (2003) Genome-wide analysis of NBS-LRR encoding genes in *Arabidopsis*. The Plant Cell Online 15, 809-834.

Mhamdi A., G. Queval, S. Chaouch, S. Vanderauwera, F. Van Breusegem & G. Noctor. (2010) Catalase function in plants: A focus on *Arabidopsis* mutants as stress-mimic models. Journal of Experimental Botany 61, 4197-4220.

Miller K.A. & P. Wagner (1994) Out of Ireland: The story of Irish emigration to America. Washington, DC., Elliot and Clark.

Mittler R. (2002) Oxidative stress, antioxidants and stress tolerance. Trends in Plant Science 7, 405-410.

Mittler R., S. Vanderauwera, M. Gollery & F. Van Breusegem. (2004) Reactive oxygen gene network of plants. Trends in Plant Science 9, 490-498.

Miura K., S.Y. Lin, M. Yano & T. Nagamine. (2002) Mapping quantitative trait loci controlling seed longevity in rice (*Oryza sativa* L.). Theoretical and Applied Genetics 104, 981-986.

Mohamed-Yasseen Y., S.A. Barringer, W.E. Splittstoesser & S. Costanza. (1994) The role of seed coats in seed viability. Botanical Review 60, 426-439.

Moise J.A., S. Han, L. Gudynaite-Savitch, D.A. Johnson & B.L.A. Miki. (2005) Seed coats: Structure, development, composition, and biotechnology. In Vitro Cellular & Developmental Biology-Plant 41, 620-644.

Moller I.M., P.E. Jensen & A. Hansson. (2007) Oxidative modifications to cellular components in plants. Annual Review of Plant Biology 58, 459-481.

Monaghan P., A. Charmantier, D.H. Nussey & R.E. Ricklefs. (2008) The evolutionary ecology of senescence. Functional Ecology 22, 371-378.

Müller K., C. Job, M. Belghazi, D. Job & G. Leubner-Metzger. (2010) Proteomics reveal tissue-specific features of the cress (*Lepidium sativum* L.) endosperm cap proteome and its hormone-induced changes during seed germination. Proteomics 10, 406-416.

Myles S., J. Peiffer, P.J. Brown, E.S. Ersoz, Z.W. Zhang, D.E. Costich & E.S. Buckler. (2009) Association mapping: Critical considerations shift from genotyping to experimental design. Plant Cell 21, 2194-2202.

Nagel M. & A. Börner. (2010) The longevity of crop seeds stored under ambient conditions. Seed Science Research 20, 1-12.

Nagel M., M. Abdur Rehman Arif, M. Rosenhauer & A. Börner (2010) Longevity of seeds - intraspecific differences in the Gatersleben genebank collections. pp 179-181 Proceedings of the 60th conference of plant breeders and seed traders association in Austria 2009. Raumberg-Gumpenstein.

Nagel M., M. Rosenhauer, E. Willner, R.J. Snowdon, W. Friedt & A. Börner. (2011) Seed longevity in oilseed rape (*Brassica napus* L.) - genetic variation and QTL mapping. Plant Genetic Resources: Characterization and Utilization 9, 260-263.

Noctor G., A.-C.M. Arisi, L. Jouanin, K.J. Kunert, H. Rennenberg & C.H. Foyer. (1998) Glutathione: Biosynthesis, metabolism and relationship to stress tolerance explored in transformed plants. Journal of Experimental Botany 49, 623-647.

Nordborg M. & D. Weigel. (2008) Next-generation genetics in plants. Nature 456, 720-723.

Nowotny M., S.A. Gaidamakov, R.J. Crouch & W. Yang. (2005) Crystal structures of RNase H bound to an RNA/DNA hybrid: Substrate specificity and metal-dependent catalysis. Cell 121, 1005-1016.

Ohme-Takagi M. & H. Shinshi. (1995) Ethylene-inducible DNA binding proteins that interact with an ethylene-responsive element. The Plant Cell Online 7, 173-182.

Oliver M.J., Z. Tuba & B.D. Mishler. (2000) The evolution of vegetative desiccation tolerance in land plants. Plant Ecology 151, 85-100.

Oracz K., H. El-Maarouf-Bouteau, I. Kranner, R. Bogatek, F. Corbineau & C. Bailly. (2009) The mechanisms involved in seed dormancy alleviation by hydrogen cyanide unravel the role of reactive oxygen species as key factors of cellular signaling during germination. Plant Physiology 150, 494-505.

Oracz K., H. El-Maarouf-Bouteau, J.M. Farrant, K. Cooper, M. Belghazi, C. Job, D. Job, F. Corbineau & C. Bailly. (2007) ROS production and protein oxidation as a novel mechanism for seed dormancy alleviation. Plant Journal 50, 452-465.

Osborne D.J. (2000) Hazards of a germination seed: Available water and the maintenance of genomic integrity. Israel Journal of Plant Sciences 48, 173-179.

Patist A. & H. Zoerb. (2005) Preservation mechanisms of trehalose in food and biosystems. Colloids and Surfaces B: Biointerfaces 40, 107-113.

Pei Z.-M., Y. Murata, G. Benning, S. Thomine, B. Klusener, G.J. Allen, E. Grill & J.I. Schroeder. (2000) Calcium channels activated by hydrogen peroxide mediate abscisic acid signalling in guard cells. Nature 406, 731-734.

Pereira E.J., A.D. Panek & E.C.A. Eleutherio. (2003) Protection against oxidation during dehydration of yeast. Cell Stress & Chaperones 8, 120-124.

Petre B., I. Major, N. Rouhier & S. Duplessis. (2011) Genome-wide analysis of eukaryote thaumatin-like proteins (TLPs) with an emphasis on poplar. BMC Plant Biology 11, 33.

Peumans W. & E. Van Damme. (1995) The role of lectins in plant defence. The Histochemical Journal 27, 253-271.

Piotrowicz-Cieslak A.I., M. Niedzielski, D.J. Michalczyk, W. Luczak & B. Adomas. (2010) Soluble carbohydrates in cereal (wheat, rye, triticale) seed after storage under accelerated ageing conditions. Acta Societatis Botanicorum Poloniae 79, 21-25.

Pourcel L., J.-M. Routaboul, V. Cheynier, L. Lepiniec & I. Debeaujon. (2007) Flavonoid oxidation in plants: From biochemical properties to physiological functions. Trends in Plant Science 12, 29-36.

Pourkheirandish M. & T. Komatsuda. (2007) The importance of barley genetics and domestication in a global perspective. Annals of Botany 100, 999-1008.

Powell A.A. (1989) The importance of genetically determined seed coat characteristics to seed quality in grain legumes. Annals of Botany 63, 169-175.

Prat H. (1952) Microcalorimetric studies on germination of cereals. Canadian Journal of Botany 30, 379-394.

Price A.H. (2006) Believe it or not, QTLs are accurate! Trends in Plant Science 11, 213-216.

Price A.L., N.J. Patterson, R.M. Plenge, M.E. Weinblatt, N.A. Shadick & D. Reich. (2006) Principal components analysis corrects for stratification in genome-wide association studies. Nature Genetics 38, 904-909.

Priestley D.A., M.B. McBride & C. Leopold. (1980) Tocopherol and organic free radical levels in soybean seeds during natural and accelerated aging. Plant Physiology 66, 715-719.

Priestley D.A., V.I. Cullinan & J. Wolfe. (1985) Differences in seed longevity at the species level. Plant Cell and Environment 8, 557-562.

Probert R., J. Adams, J. Coneybeer, A. Crawford & F. Hay. (2007) Seed quality for conservation is critically affected by pre-storage factors. Australian Journal of Botany 55, 326-335.

Pukacka S. & E. Ratajczak. (2007) Age-related biochemical changes during storage of beech (*Fagus sylvatica* L.) seeds. Seed Science Research 17, 45-53.

Queval G., E. Issakidis-Bourguet, F.A. Hoeberichts, M. Vandorpe, B. Gakiere, H. Vanacker, M. Miginiac-Maslow, F. Van Breusegem & G. Noctor. (2007) Conditional oxidative stress responses in the *Arabidopsis* photorespiratory mutant cat2 demonstrate that redox state is a key modulator of daylength-dependent gene expression, and define photoperiod as a crucial factor in the regulation of H_2O_2-induced cell death. Plant Journal 52, 640-657.

Rajjou L. & I. Debeaujon. (2008) Seed longevity: Survival and maintenance of high germination ability of dry seeds. Comptes Rendus Biologies 331, 796-805.

Rajjou L., Y. Lovigny, S.P.C. Groot, M. Belghaz, C. Job & D. Job. (2008) Proteome-wide characterization of seed aging in *Arabidopsis*: A comparison between artificial and natural aging protocols. Plant Physiology 148, 620-641.

Ramanatha Rao V. & T. Hodgkin. (2002) Genetic diversity and conservation and utilization of plant genetic resources. Plant Cell, Tissue and Organ Culture 68, 1-19.

Ratajczak E. & S. Pukacka. (2005) Decrease in beech (*Fagus sylvatica*) seed viability caused by temperature and humidity conditions as related to membrane damage and lipid composition. Acta Physiologiae Plantarum 27, 3-12.

Rauber R. (1987) Überdauern von frischgereiften Wintergerstenkörnern (*Hordeum vulgare* L.) in verschiedenen Bodentiefen. Angewandte Botanik 61, 325-335.

Rauser W.E., R. Schupp & H. Rennenberg. (1991) Cysteine, g-glutamylcysteine, and glutathione levels in maize seedlings : Distribution and translocation in normal and cadmium-exposed plants. Plant Physiology 97, 128-138.

Revilla P., P. Velasco, R.A. Malvar, M.E. Cartea & A. Ordas. (2006) Variability among maize (*Zea mays* L.) inbred lines for seed longevity. Genetic Resources and Crop Evolution 53, 771-777.

Revilla P., A. Butron, V.M. Rodriguez, R.A. Malvar & A. Ordas. (2009) Identification of genes related to germination in aged maize seed by screening natural variability. Journal of Experimental Botany 60, 4151-4157.

Rhazi L., R. Cazalis, E. Lemelin & T. Aussenac. (2003) Changes in the glutathione thiol-disulfide status during wheat grain development. Plant Physiology and Biochemistry 41, 895-902.

Riedl S.J. & Y. Shi. (2004) Molecular mechanisms of caspase regulation during apoptosis. Nat Rev Mol Cell Biol 5, 897-907.

Roberts E.H. (1960) The viability of cereal seed in relation to temperature and moisture: With eight figures in the text. Annals of Botany 24, 12-31.

Roberts E.H. & R.H. Ellis. (1977) Prediction of seed longevity at sub-zero temperatures and genetic resources conservation. Nature 268, 431-433.

Roos E.E. & D.A. Davidson. (1992) Record longevities of vegetable seeds in storage. Hortscience 27, 393-396.

Rostoks N., L. Ramsay, K. MacKenzie, L. Cardle, P.R. Bhat, M.L. Roose, J.T. Svensson, N. Stein, R.K. Varshney, D.F. Marshall, A. Graner, T.J. Close & R. Waugh. (2006) Recent history of artificial outcrossing facilitates whole-genome association mapping in elite inbred crop varieties. Proceedings of the National Academy of Science 103, 18656-18661

Saitoh M., H. Nishitoh, M. Fujii, K. Takeda, K. Tobiume, Y. Sawada, M. Kawabata, K. Miyazono & H. Ichijo. (1998) Mammalian thioredoxin is a direct inhibitor of apoptosis signal-regulating kinase (ASK) 1. EMBO J 17, 2596-2606.

Sakuma Y., Q. Liu, J.G. Dubouzet, H. Abe, K. Shinozaki & K. Yamaguchi-Shinozaki. (2002) DNA-binding specificity of the ERF/AP2 domain of *Arabidopsis* DREBs, transcription factors involved in dehydration- and cold-inducible gene expression. Biochemical and Biophysical Research Communications 290, 998-1009.

Sallon S., E. Solowey, Y. Cohen, R. Korchinsky, M. Egli, I. Woodhatch, O. Simchoni & M. Kislev. (2008) Germination, genetics, and growth of an ancient date seed. Science 320, 1464-1464.

Sattler S.E., L.U. Gilliland, M. Magallanes-Lundback, M. Pollard & D. DellaPenna. (2004) Vitamin E is essential for seed longevity, and for preventing lipid peroxidation during germination. Plant Cell 16, 1419-1432.

Schafer F.Q. & G.R. Buettner. (2001) Redox environment of the cell as viewed through the redox state of the glutathione disulfide/glutathione couple. Free Radical Biology and Medicine 30, 1191-1212.

Schulte D., T.J. Close, A. Graner, P. Langridge, T. Matsumoto, G. Muehlbauer, K. Sato, A.H. Schulman, R. Waugh, R.P. Wise & N. Stein. (2009) The international barley sequencing consortium - at the threshold of efficient access to the barley genome. Plant Physiology 149, 142-147.

Schwember A.R. & K.J. Bradford. (2010) Quantitative trait loci associated with longevity of lettuce seeds under conventional and controlled deterioration storage conditions. Journal of Experimental Botany 61, 4423-4436.

Seal C.E., R. Zammit, P. Scott, T.J. Flowers & I. Kranner. (2010) Glutathione half-cell reduction potential and a-tocopherol as viability markers during the prolonged storage of *Suaeda maritima* seeds. Seed Science Research 20, 47-53.

Sebastiani P., N. Solovieff, A. Puca, S.W. Hartley, E. Melista, S. Andersen, D.A. Dworkis, J.B. Wilk, R.H. Myers, M.H. Steinberg, M. Montano, C.T. Baldwin & T.T. Perls. (2010) Genetic signatures of exceptional longevity in humans. Science.

Seki M., M. Narusaka, H. Abe, M. Kasuga, K. Yamaguchi-Shinozaki, P. Carninci, Y. Hayashizaki & K. Shinozaki. (2001) Monitoring the expression pattern of 1300 *Arabidopsis* genes under drought and cold stresses by using a full-length cDNA microarray. The Plant Cell Online 13, 61-72.

Sen S. & D.J. Osborne. (1974) Germination of rye embryos following hydration-dehydration treatments - enhancement of protein and RNA-synthesis and earlier induction of DNA-replication. Journal of Experimental Botany 25, 1010-1019.

Sen S. & D.J. Osborne. (1977) Decline in ribonucleic-acid and protein-synthesis with loss of viability during early hours of imbibition of rye (*Secale cereale* L.) embryos. Biochemical Journal 166, 33-38.

Senaratna T., B.D. McKersie & A. Borochov. (1987) Desiccation and free radical mediated changes in plant membranes. Journal of Experimental Botany 38, 2005-2014.

Shen-Miller J. (2002) Sacred lotus, the long-living fruits of China Antique. Seed Science Research 12, 131-143.

Shen-Miller J., M.B. Mudgett, J.W. Schopf, S. Clarke & R. Berger. (1995) Exceptional seed longevity and robust growth - ancient sacred lotus from China. American Journal of Botany 82, 1367-1380.

Shinozaki K. & K. Yamaguchi-Shinozaki. (2000) Molecular responses to dehydration and low temperature: Differences and cross-talk between two stress signaling pathways. Current Opinion in Plant Biology 3, 217-223.

Singh R.K., R.K. Raipuria, V.S. Bhatia, A. Rani, Pushpendra, S.M. Husain, D. Chauhan, G.S. Chauhan & T. Mohapatra. (2008) SSR markers associated with seed longevity in soybean. Seed Science and Technology 36, 162-167.

Smirnoff N. (2010) Tocochromanols: Rancid lipids, seed longevity, and beyond. Proceedings of the National Academy of Sciences 107, 17857-17858.

Sreenivasulu N., A. Graner & U. Wobus. (2008) Barley genomics: An overview. International Journal of Plant Genomics 2008, 1-13.

Sreenivasulu N., V. Radchuk, M. Strickert, O. Miersch, W. Weschke & U. Wobus. (2006) Gene expression patterns reveal tissue-specific signaling networks controlling programmed cell death and ABA-regulated maturation in developing barley seeds. The Plant Journal 47, 310-327.

Steadman K.J., H.W. Pritchard & P.M. Dey. (1996) Tissue-specific soluble sugars in seeds as indicators of storage category. Annals of Botany 77, 667-674.

Stein N., M. Prasad, U. Scholz, T. Thiel, H.N. Zhang, M. Wolf, R. Kota, R.K. Varshney, D. Perovic, I. Grosse & A. Graner. (2007) A 1,000-loci transcript map of the barley genome: New anchoring points for integrative grass genomics. Theoretical and Applied Genetics 114, 823-839.

Steiner A.M. & P. Ruckenbauer. (1995) Germination of 110-year-old cereal and weed seeds, the Vienna Sample of 1877. Verification of effective ultra-dry storage at ambient temperature. Seed Science Research 5, 195-199.

Stracke S., G. Haseneyer, J.B. Veyrieras, H.H. Geiger, S. Sauer, A. Graner & H.P. Piepho. (2009) Association mapping reveals gene action and interactions in the determination of flowering time in barley. Theoretical and Applied Genetics 118, 259-273.

Strelec I., Z. Ugarcic-Hardi & M. Hlevnjak. (2008) Accumulation of Amadori and Maillard products in wheat seeds aged under different storage conditions. Croatica Chemica Acta 81, 131-137.

Sugliani M., L. Rajjou, E.J.M. Clerkx, M. Koornneef & W.J.J. Soppe. (2009) Natural modifiers of seed longevity in the *Arabidopsis* mutants abscisic acid insensitive3-5 (*abi3-5*) and leafy cotyledon1-3 (*lec1-3*). New Phytologist 184, 898-908.

Sun S., J.-P. Yu, F. Chen, T.-J. Zhao, X.-H. Fang, Y.-Q. Li & S.-F. Sui. (2008) TINY, a dehydration-responsive element (DRE)-binding protein-like transcription factor connecting the DRE- and Ethylene-responsive element-mediated signaling pathways in *Arabidopsis*. Journal of Biological Chemistry 283, 6261-6271.

Sun W.Q. (1997) Glassy state and storage stability: The WLF kinetics of seed viability loss at $T > T_g$ and the plasticization effect of water on storage stability. Annals of Botany 79, 291-297.

Sun W.Q. & A.C. Leopold. (1995) The Maillard reaction and oxidative stress during aging of soybean seeds. Physiologia Plantarum 94, 94-104.

Sun W.Q. & A.C. Leopold. (1997) Cytoplasmic vitrification acid survival of anhydrobiotic organisms. Comparative Biochemistry and Physiology a-Physiology 117, 327-333.

Sung J.M. & T.L. Jeng. (1994) Lipid-peroxidation and peroxide-scavenging enzymes associated with accelerated aging of peanut seed. Physiologia Plantarum 91, 51-55.

Swanson S.J., P.C. Bethke & R.L. Jones. (1998) Barley aleurone cells contain two types of vacuoles: Characterization of lytic organelles by use of fluorescent probes. Plant Cell 10, 685-698.

Szick K., M. Springer & J. Bailey-Serres. (1998) Evolutionary analyses of the 12-kDa acidic ribosomal P-proteins reveal a distinct protein of higher plant ribosomes. Proceedings of the National Academy of Sciences 95, 2378-2383.

Tabien R.E., S.O.P.B. Samonte, C.L. Harper & P.M. Frank. (2008) Variations in farmer's seed and genebank accessions of Milagro Filipino rice (*Oryza sativa* L.) released in Mexico in 1967. Sabrao Journal of Breeding and Genetics 40, 37-47.

Tammela P., M. Nygren, I. Laakso, A. Hopia, H. Vuorela & R. Hiltunen. (2003) Volatile compound analysis of ageing *Pinus sylvestris* L. (Scots pine) seeds. Flavour and Fragrance Journal 18, 290-295.

Tanksley S.D. & S.R. McCouch. (1997) Seed banks and molecular maps: Unlocking genetic potential from the wild. Science 277, 1063-1066.

Terskikh V., J. Feurtado, C. Ren, S. Abrams & A. Kermode. (2005a) Water uptake and oil distribution during imbibition of seeds of western white pine (*Pinus monticola* Dougl. ex D. Don) monitored *in vivo* using magnetic resonance imaging. Planta 221, 17-27.

Terskikh V.V., J.A. Feurtado, S. Borchardt, M. Giblin, S.R. Abrams & A.R. Kermode. (2005b) *In vivo*13C NMR metabolite profiling: Potential for understanding and assessing conifer seed quality. Journal of Experimental Botany 56, 2253-2265.

Tesnier K., H.M. Strookman-Donkers, J.G. Van Pijlen, A.H.M. Van der Geest, R.J. Bino & S.P.C. Groot. (2002) A controlled deterioration test for *Arabidopsis thaliana* reveals genetic variation in seed quality. Seed Science and Technology 30, 149-165.

Tobias D., M. Manoharan, C. Pritsch & L. Dahleen. (2007) Co-bombardment, integration and expression of rice chitinase and thaumatin-like protein genes in barley (*Hordeum vulgare* cv. Conlon). Plant Cell Reports 26, 631-639.

Tommasini R., E. Martinoia, E. Grill, K.J. Dietz & N. Amrhein. (1993) Transport of oxidized glutathione into barley vacuoles - evidence for the involvement of the glutathione *S*-conjugate ATPase. Zeitschrift Fur Naturforschung C-a Journal of Biosciences 48, 867-871.

Toyooka K., T. Okamoto & T. Minamikawa. (2001) Cotyledon cells of *Vigna mungo* seedlings use at least two distinct autophagic machineries for degradation of starch granules and cellular components. The Journal of Cell Biology 154, 973-982.

Vanderauwera S., N. Suzuki, G. Miller, B. van de Cotte, S. Morsa, J.-L. Ravanat, A. Hegie, C. Triantaphylidès, V. Shulaev, M.C.E. Van Montagu, F. Van Breusegem & R. Mittler. (2011) Extranuclear protection of chromosomal DNA from oxidative stress. Proceedings of the National Academy of Sciences 108, 1711-1716.

Varier A., A. Kuriakose Vari & M. Dadlani. (2010) The subcellular basis of seed priming. Current Science 99, 450-456.

Varshney R.K., M. Baum, P. Guo, S. Grando, S. Ceccarelli & A. Graner. (2010) Features of SNP and SSR diversity in a set of ICARDA barley germplasm collection. Molecular Breeding 26, 229-242.

Vavilov N.I. (1926) Studies on the origin of cultivated plants. Bulletin of Applied Botany, of Genetics and Plant Breeding 16, 1-248.

Vavilov N.I. (2009) Geographical regularities in the distribution of the genes of cultivated plants. Comparative Cytogenetics 3, 71-78.

Velazhahan R. & S. Muthukrishnan. (2003) Transgenic tobacco plants constitutively overexpressing a rice thaumatin-like protein (PR-5) show enhanced resistance to *Alternaria alternata*. Biologia Plantarum 47, 347-354.

Vertucci C.W. & E.E. Roos. (1990) Theoretical basis of protocols for seed storage. Plant Physiology 94, 1019-1023.

Vertucci C.W. & J.M. Farrant. (1995) Acquisition and loss of desiccation tolerance. pp 237–272 In: J. Kigel & G. Galili (Eds) Seed development and germination. New York, Marcel Dekker Inc.

Vijg J. & Y. Suh. (2005) Genetics of longevity and aging. Annual Review of Medicine 56, 193-212.

Walters C. (1998) Understanding the mechanisms and kinetics of seed aging. Seed Science Research 8, 223-244.

Walters C. (2004) Principles for preserving germplasm in gene banks In: E.O. Guerrant, K. Haven, M. Maunder & T.H. Raven (Eds) *Ex situ* plant conservation: Supporting species survival in the wild. Washington, D.C., Island Press.

Walters C., L. Wheeler & P.C. Stanwood. (2004) Longevity of cryogenically stored seeds. Cryobiology 48, 229-244.

Walters C., L.M. Wheeler & J.M. Grotenhuis. (2005a) Longevity of seeds stored in a genebank: Species characteristics. Seed Science Research 15, 1-20.

Walters C., L.M. Hill & L.J. Wheeler. (2005b) Dying while dry: Kinetics and mechanisms of deterioration in desiccated organisms. Integrative and Comparative Biology 45, 751-758.

Waterworth W.M., C. Altun, S.J. Armstrong, N. Roberts, P.J. Dean, K. Young, C.F. Weil, C.M. Bray & C.E. West. (2007) NBS1 is involved in DNA repair and plays a synergistic role with ATM in mediating meiotic homologous recombination in plants. Plant Journal 52, 41-52.

Waugh R., J.L. Jannink, G.J. Muehlbauer & L. Ramsay. (2009) The emergence of whole genome association scans in barley. Current Opinion in Plant Biology 12, 218-222.

Weil C.F. (2009) TILLING in grass species. Plant Physiology 149, 158-164.

Wenzl P., J. Carling, D. Kudrna, D. Jaccoud, E. Huttner, A. Kleinhofs & A. Kilian. (2004) Diversity Arrays Technology (DArT) for whole-genome profiling of barley. Proceedings of the National Academy of Sciences of the United States of America 101, 9915-9920.

Wenzl P., H.B. Li, J. Carling, M.X. Zhou, H. Raman, E. Paul, P. Hearnden, C. Maier, L. Xia, V. Caig, J. Ovesna, M. Cakir, D. Poulsen, J.P. Wang, R. Raman, K.P. Smith, G.J. Muehlbauer, K.J. Chalmers, A. Kleinhofs, E. Huttner & A. Kilian (2006) A high-density consensus map of barley linking DArT markers to SSR, RFLP and STS loci and agricultural traits. BMC Genomics 7, -.

Williams K.J. (2003) The molecular genetics of disease resistance in barley. Australian Journal of Agricultural Research 54, 1065-1079.

Williams R., A. Hirsh, H. Meryman & T. Takahashi. (1993) The high-order kinetics of cytolysis in stressed red cells. Journal of Thermal Analysis and Calorimetry 40, 857-862.

Winkel-Shirley B. (2002) Biosynthesis of flavonoids and effects of stress. Current Opinion in Plant Biology 5, 218-223.

Wobus U. & I. Schubert. (2002) Science and politics: Hans Stubbe and the Institute of Plant Genetics and Crop Plant Research at Gatersleben. Trends in Plant Science 7, 418-420.

Wolkers W.F., M.G.v. Kilsdonk & F.A. Hoekstra. (1998) Dehydration-induced conformational changes of poly-L-lysine as influenced by drying rate and carbohydrates. Biochimica et Biophysica Acta (BBA) - Molecular and Cell Biology of Lipids 1425, 127-136.

Wolkers W.F., A.E. Oliver, F. Tablin & J.H. Crowe. (2004) A Fourier-transform infrared spectroscopy study of sugar glasses. Carbohydrate Research 339, 1077-1085.

Wolkers W.F., S. McCready, W.F. Brandt, G.G. Lindsey & F.A. Hoekstra. (2001) Isolation and characterization of a D-7 LEA protein from pollen that stabilizes glasses in vitro. Biochimica et Biophysica Acta (BBA) - Protein Structure and Molecular Enzymology 1544, 196-206.

Xue Y., S.Q. Zhang, Q.H. Yao, R.H. Peng, A.S. Xiong, X. Li, W.M. Zhu, Y.Y. Zhu & D.S. Zha. (2008) Identification of quantitative trait loci for seed storability in rice (*Oryza sativa* L.). Euphytica 164, 739-744.

Yan Y., F.A. Popp & G.M. Rothe. (2003) Correlation between germination capacity and biophoton emission of barley seeds (*Hordeum vulgare* L.). Seed Science and Technology 31, 249-258.

Yeh Y.M. & J.M. Sung. (2008) Priming slows deterioration of artificially aged bitter gourd seeds by enhancing anti-oxidative activities. Seed Science and Technology 36, 350-359.

Yu J., G. Pressoir, W.H. Briggs, I. Vroh Bi, M. Yamasaki, J.F. Doebley, M.D. McMullen, B.S. Gaut, D.M. Nielsen, J.B. Holland, S. Kresovich & E.S. Buckler. (2006) A unified mixed-model method for association mapping that accounts for multiple levels of relatedness. Nature Genetics 38, 203-208.

Zakharov I. (2005) Nikolai I Vavilov (1887–1943). Journal of biosciences 30, 299-301.

Zeng D.L., L.B. Guo, Y.B. Xu, K. Yasukumi, L.H. Zhu & Q. Qian. (2006) QTL analysis of seed storability in rice. Plant Breeding 125, 57-60.

Zhou Y., H. Cai, J. Xiao, X. Li, Q. Zhang & X. Lian. (2009) Over-expression of aspartate aminotransferase genes in rice resulted in altered nitrogen metabolism and increased amino acid content in seeds. Theoretical and Applied Genetics 118, 1381-1390.

Zhu T., P. Budworth, B. Han, D. Brown, H.-S. Chang, G. Zou & X. Wang. (2001) Toward elucidating the global gene expression patternsof developing *Arabidopsis*: Parallel analysis of 8.300 genes by a high-density oligonucleotide probe array. Plant Physiology and Biochemistry 39, 221-242.

Zohary D. & M. Hopf (2000) Domestication of plants in the old world. New York, Oxford University Press.

Acknowledgements/ Danksagung

Vor ungefähr 10 Jahren, nach der Beendigung meiner Schulzeit sollte sich ursprünglich ein Luft- und Raumfahrttechnik Studium oder eine Pilotenausbildung anschließen, aber die Entscheidung ist letztendlich auf etwas Grundsolides gefallen, die Landwirtschaft. Auf das ernüchternde Grundstudium folgten spannende Praktika in der Tier– und Pflanzenwelt, die mich bis in weit entlegene Gebiete brachten und dazu führten, dass ich mich letztendlich den Tieren privat und den Pflanzen beruflich widmete.

Vielen Dank Herr Prof. Rolf Rauber, dass Sie mir ein Stück diesen Weg bereitet und durch interessante Vorlesungen, Exkursionen und Praktika während der Uni bereichert haben. Ihre Samensammlung ist immer noch in meinen Besitz, wobei sich diese arbeitstechnisch um ein Tausendfaches erweitert hat. Vielen Dank Ihnen und Ihrer Frau für die Mühen während der Promotion um auch einer externen Studentin dieses Studium zu ermöglichen.

Die Begeisterung für Wissenschaft wurde dann am IPK zur ansteckenden Krankheit. Vielen Dank, Herr Dr. Andreas Börner, dass Sie der Überträger waren und mir Konferenzen, Workshops, Auslandsaufenthalte und den Kontakt mit internationalen Forschern erst ermöglicht haben. Glaubte ich doch zu Beginn, man wüsste schon alles über diese kleinen ‚unspannenden' Samen, so musste ich doch eine faszinierende Welt in Miniaturformat (Samen) entdecken, die bisher nur wenige Geheimnisse preisgegeben hat. Zwar sind wir von einer patentierten neuen Methode zur Keimfähigkeitsprüfung weit entfernt und auch der Nature Artikel muss auf sich warten, aber einige Ergebnisse haben uns diesen Dingen vielleicht ein kleines Stück näher gebracht.

One of the most exciting experiences was it to get to know another seedbank in another country including enthusiasm scientists who also try to unlock the secrets of seed ageing. Dear Ilse Kranner and dear Hugh Pritchard, thank you very much for the great opportunities to join you and your team. I did learn a lot about the scientific way of thinking, paper writing and sharing chocolate cake after dinner. Thank you, Charlotte Seal and Louise Colville, for your great help and your patience in the lab. It was probably a challenge but you could somehow manage to mutate a farmer girl to a lab mouse (unfortunately not without losses).

Eine Vielzahl der Ergebnisse wäre kaum zu Stande gekommen, hätte ich nicht Euch, Sibylle Pistrick, Anita Winger und Stefanie Thumm, an meiner Seite gehabt. Vielen Dank, dass ihr bei jeder Untat selbst innerhalb kürzester Zeit mit gesät, bonitiert, geerntet, gedroschen, gereinigt, gewogen, gezählt, gemessen, gepulvert und geschleppt habt, auch wenn es 800 Vermehrungstüten (2-5kg) waren, die nach drei Umlagerungen auf den Boden kamen mit der festen Gewissheit irgendjemand muss sie von dort wieder runterholen. Vielen Dank liebe Sibylle, für die vielen Wochenendeinsätze, der schnellen Organisation von Informationen und, dass Du mich von den allerdümmsten Ideen abgehalten hast. Auch an Euch, Annette Marlow und Renate Voß, vielen lieben Dank für Ideen, Rat und Tat bei der Entwicklung von Experimenten und anderer Schandtaten.

Die herzliche Aufnahme in der Arbeitsgruppe und die tolle Unterstützung aller Sortimentsgruppen, die ich zu jeder Zeit bekommen habe, möchte ich natürlich nicht vergessen. Vielen, vielen Dank. Insbesondere geht auch ein großes Dankschön an Dich

Michael Grau und Dein Team für die wichtigen Informationen bei den Feldvermehrungen und die immer kurzfristig zur Verfügung gestellten Hilfeleistungen, ebenso vielen Dank Frau Bärbel Schmidt für die Unterstützung bei den Bohnenanbauten. Und natürlich möchte ich mich auch bei Ihnen Herr Peter Schreiber und Herr Jürgen Marlow und Ihren Teams von emsig arbeitenden Kollegen bedanken, ohne die meine Felder ein Desaster geworden wären. Einem weiterem großem Dank bin ich auch meinen Studenten verpflichtet, die maßgeblich an den Ergebnissen mitgewirkt haben. Vielen lieben Dank Anne-Kathrin Behrens, Arne Brathuhn, Franziska Scharkowski, Heike Vogel, Judith Jäger, Jutta Scheurenberg, Maria Rosenhauer, Mariann Börner, Nadja Borasca und Sandra Minnich. Für experimentelle Zusammenarbeiten und analytische Unterstützung möchte ich mich auch bei Euch, Evelin Willner und Dr. Ulrike Lohwasser bedanken. Und ein sehr großes Dankschön geht an Dich Kerstin Neumann, für die stetige Bereitschaft mir mit Rat und Tat beiseite zu stehen. Ohne Deine Vorarbeiten wäre meine Arbeit nie so schnell fertig geworden.

Für die kurzfristige Bereitschaft sich durch diese Arbeit zu kämpfen, möchte ich mich auch bei Euch Christine Zanke und Kerstin Diekmann bedanken. Es war mir wirklich eine riesen Hilfe auf den letzten Metern. Ebenso vielen Dank Gudrun Schütze und Roswitha Selbig für die stetige Unterstützung in diversen Belangen.

Zum Schluss möchte ich mich nun bei den Menschen aus tiefsten Herzen bedanken, ohne die das alles überhaupt nicht möglich gewesen wäre: bei meinen Eltern, Ilona und Wolfhard Nagel, und meinem Freund Martin Henning. Danke schön, dass ihr an mich geglaubt, mich unterstützt, motiviert und sämtliche Eskapaden mitgemacht habt. Weiterhin vielen Dank an Christina und Reinhard Henning und meine Freunde für das erfolgreiche Aufladen meines Akkus falls dieser einmal wieder erschöpft war.

.

Curriculum vitae

Name:	Manuela Nagel
Date of Birth:	04th October 1981
Place of Birth:	Quedlinburg, Saxony-Anhalt, Germany

Education

Since Oct 2007	The University of Göttingen Ph.D. Program for Agricultural Sciences in Göttingen
2000 – 2007	The University of Göttingen Master of Science in Agricultural Sciences Bachelor of Science in Agribusiness
2002 – 2003	State Office for Agriculture and Horticulture, Saxony-Anhalt Internship including examination
1992 – 2000	Wolterstorff-Gymnasium Ballenstedt

Experience

Since 2006	Scientist at Genebank Department, IPK Gatersleben
2009, 2010	Royal Botanical Gardens KEW, Millennium Seedbank, UK
2005 – 2006	Internship on a dairy farm in Rakaia, New Zealand
2003	Internship on an agricultural cooperative in Hedersleben
2002	Nordsaat Saatzucht GmbH (Plant Breeding Company)

Awards

June 2010	ISTA Congress Köln Seed Symposium Award 2010 for the oral presentation
Feb 2010	Conference of the German Society for Quality Research, Berlin Poster Award 2010
June 2009	Plant Science Student Conference 2009 in Halle Second price for the oral presentation

Publications

Nagel M. & A. Börner. (2008) Inter- und intraspezifische Variabilität der Lebensfähigkeit von Saatgut der bundeszentralen *ex situ* Genbank in Gatersleben. Mitteilungen der Gesellschaft für Pflanzenbauwissenschaften Band 20, Vorträge für Pflanzenzüchtung Heft 77, 49-50.

Nagel M., S. Pistrick & A. Börner. (2008) Langlebigkeit von Saatgut in der *ex situ* Genbank in Gatersleben. 58. Tagung der Vereinigung der Pflanzenzüchter und Saatgutkaufleute Österreichs 2007, Raumberg-Gumpenstein, Austria. 59-62.

Werner C., M. Nagel & M. Wicke. (2008) Influence of genetic, sex and lairage duration on the incidence of hemorrhages in different broiler muscles after slaughter. Fleischwirtschaft International 2, 77-80.

Nagel M. & A. Börner. (2009) Genetische Hintergründe zur Langlebigkeit von Gerste. Mitteilungen der Gesellschaft für Pflanzenbauwissenschaften Band 21, 229-230.

Nagel M., H. Vogel, S. Landjeva, G. Buck-Sorlin, U. Lohwasser, U. Scholz & A. Börner. (2009) Seed conservation in *ex-situ* genebanks- genetic studies on longevity in barley. Euphytica 170,1-10.

Nagel M. & A. Börner. (2010) The longevity of crop seeds stored under ambient conditions. Seed Science Research 20, 1-12.

Nagel M., I. Kranner & A. Börner. (2010) Langlebigkeit von Getreidesamen und deren genetische Ursachen. Berichte der Gesellschaft für Pflanzenbauwissenschaften Band 5, 7-10

Nagel M., M. A. Rehman Arif, I. Kranner, A. Fiedler, H. Schulz & A. Börner. (2010) Physiologische und genetische Aspekte zur Qualität von langzeitgelagertem Weizensaatgut. 45. DGQ Vortragstagung, Berlin-Dahlem, 53-54

Nagel M., M. A. Rehman Arif, M. Rosenhauer & A. Börner. (2010) Longevity of seeds - intraspecific differences in the Gatersleben genebank collections. 60. Tagung der Vereinigung der Pflanzenzüchter und Saatgutkaufleute Österreichs 2009, Raumberg-Gumpenstein, Austria, 179-181.

Nagel, M., M. Rosenhauer, E. Willner, R.J. Snowdon, W. Friedt & A. Börner. (2011) Seed longevity in oilseed rape (*Brassica napus* L.) - genetic variation and QTL mapping. Plant Genetic Resources: Characterization and Utilization 9, 260-263.

Rehman Arif M. A., M. Nagel, K. Neumann, B. Kobiljski, U. Lohwasser & A. Börner. (2011) Genetic studies of seed longevity in hexaploid wheat exploiting segregation and association mapping approaches. Euphytica, accepted

Oral presentations at conferences

Nov 2007   Pflanzenzüchtertagung Raumberg-Gumpenstein (Austria)

July 2008   Plant Science Student Conference, Gatersleben

July 2008   ISSS Conference on Seed Biology, Olsztyn (Poland)

May 2009   Eucarpia Conference – Genetic Resources Section, Ljubljana (Slovenia)

June 2009   Plant Science Student Conference, Halle

Feb 2010   Arbeitstagung der AG Saatgut und Sortenwesen, Gatersleben

May 2010   IHAR Workshop, Radzikow (Poland)

June 2010   ISTA Congress Köln

Apr 2011   ISSS Conference, Salvador (Brasil) – two presentations

Poster presentations at conferences

Oct 2008   Gemeinsame Tagung der Gesellschaft für Pflanzenbauwissenschaften und der Gesellschaft für Pflanzenzüchtung, Göttingen

Sep 2009   Jahrestagung der Gesellschaft für Pflanzenbauwissenschaften, Halle

Sep 2009   Botanikertagung, Leipzig

Nov 2009   Pflanzenzüchtertagung, Raumberg-Gumpenstein (Austria)

Mar 2010   Conference of the German Society for Quality Research, Berlin

Apr 2010   International Symposium on Genomics of Plant Genetic Resources, Bologna (Italy)

June 2010   Plant Science Student Conference 2010, Gatersleben

June 2010   ISTA Congress, Köln

Apr 2011   ISSS Conference, Salvador (Brasil)